AF531440

STRESS AND ATTITUDE OF WOMEN TEACHERS

(WORKING WITH NORMAL AND SPECIAL CHILDREN)

STRESS AND ATTITUDE OF WOMEN TEACHERS

(WORKING WITH NORMAL AND SPECIAL CHILDREN)

By

S. M. RAJESWARI
M.A., M.Phil. (Psychology)
Lecturer in Psychology
Chennai, Tamil Nadu

Editors

Dr. T. SANTHANAM
M.A., B.Ed. (MR), M.Phil., Ph.D. (Psychology)
Psychologist and Head of Office
Vocational Rehabilitation Centre for Handicapped
Guindy, Chennai – 600 032

Dr. B. PRASAD BABU
M.A., B.Ed., M.Phil., Ph.D. (Psychology)
Rehabilitation Counsellor
Rural Rehabilitation Extension Centre for Handicapped
Thiruvallur, Chennai – 602 001

Dr. DIGUMARTI BHASKARA RAO
M.Sc., M.A., M.A., M.Ed., Ph.D. (Education)
Reader and Research Director
R.V.R. College of Education
Guntur – 522 006, Andhra Pradesh
digumartibhaskararao@rediffmail.com

DISCOVERY PUBLISHING HOUSE PVT. LTD.
NEW DELHI-110 002

First Published-2008
Reprint: 2011

ISBN 978-81-8356-324-6

Published by:

DISCOVERY PUBLISHING HOUSE PVT. LTD.
4831/24, Ansari Road, Prahlad Street
Darya Ganj, New Delhi-110002 (India)
Phone: 23279245 • Fax: 91-11-23253475
***E-mail:* dphbooks@rediffmail.com • dphtemp@indiatimes.com**
***web:* www.discoverypublishinghouse.com**

Printed At :- Dynamic Printers,Delhi

Dedicated
To
Legendary Educator

Dr. MARLOW EDIGER

Professor Emeritus in Education
Truman State University
United States of America
in recognition of his
excellent services to education

PREFACE

Stress is a multidimensional and a complex phenomenon, which is influenced by personal and situational factors. Specifically, stress in teaching is a multi level phenomenon that results in unpleasant negative emotions such as anger, frustration, depression, etc. Attitude is a generalized feeling or evaluative reaction to a particular object or profession, determined through psychological or socio-cultural factors.

The present study is conducted to assess the level of stress and attitude of women teachers working with normal and special children. This study reveals that the stress of women teachers as a whole is of moderate level, but those of teachers who are handling mentally challenged children is found to be relatively high compared with that of teachers handling normal children. The attitude level of the teachers is found to be moderate and there again the teachers working with mentally challenged are reporting relatively low level of attitude towards their profession. The teachers handling hearing impaired children are found to have relatively low level of stress and high attitude towards their profession. Teachers of age group of below forty years and of experience below fifteen years are found to have more stress and low attitude towards their profession. Moderately high level of stress and low level of attitude is found in the teachers working in private management schools.

This book will be of great use to psychologists, special educators and teachers along with parents and administrators in performing their jobs judiciously.

Sri Sai Soudha
D-43, S.V.N. Colony
Guntur 522006
India

Dr. D. Bhaskara Rao

CONTENTS

Preface

1. **Introduction** 1-40
   - Stress
     - Symptoms of stress
     - Health consequences of stress
     - Psychological effects of stress
   - Attitude
     - •Attitude formation
     - Determinants of attitude
     - Attitude change
   - Teaching as a profession
     - Role and responsibility of teacher
     - Basis of effective teacher
   - Special education
   - Stress in teaching
   - Causal factors of teachers stress
   - Stress in special education
   - Attitude on teaching
   - Need for the study
2. **Review of Related Research** 41-50
3. **Methodology of Research** 51-60
   - Objectives
   - Hypotheses
   - Selection of Tools
   - Selection of Sample
   - Data analysis

4. **Results and Discussion** **61-84**
Stress
Attitude
Stress and age
Attitude and age
Stress and experience
Attitude and experience
Stress and different management schools
Attitude and different management schools

5. **Summary and Conclusions** **85-88**

Bibliography *89-119*

Index *121*

1

INTRODUCTION

Education is an enterprising activity, carving the society from the ancient times. It attempts to develop the personality of an individual and then prepares him for membership in a society. Education is the modification of behaviour of an individual for a healthy social adjustment in the society. It has grown complex and complicated due to the modern day technological development, problems of life and society. The teacher plays a vital röle in spreading education and building up a healthy society. Teacher plays an important role in the cognitive, social, intellectual and emotional development of children. In education, each instance of learning involves the strengthening of a learner's tendency to respond in a certain way to a given set of circumstances. The teacher plays an important role in understanding and encouraging the individual, when he has a high degree of readiness to learn a specific response, to guide him in the response and to reinforce his natural satisfaction in his own success. The teachers are artists at recognizing, encouraging, and developing the normal desires of children to understand and make intelligent use of things that appear to concern them.

Among children, some need special efforts focused on them. Thus, special educators are the teachers, who concentrate on the needs and development of these children. The special education teachers work with children with specific learning disabilities, mental retardation, speech, hearing or visual impairment, serious emotional disturbance etc. They are involved in a student's behavioral as well as academic development.

Retention in the profession is an important aspect for teachers than other professionals. They can resent in the profession, only when their job is with less or no stress. A negative attitude towards the profession or a stressful condition may lead to some deteriorating effects to their personal and as a whole educational development of the country. Thus teachers or special educators, who build the foundation for manhood and society, should have a positive attitude on their profession and the field of education.

STRESS

Stress is a complex phenomenon. It is a very subjective experience. The term "stress" refers to an internal state, which results from demanding, frustrating or unsatisfying conditions. A certain level of stress is unavoidable. In fact, an acceptable level of stress can serve as a stimulus to enhance an individual's performance. However, when the level of stress is such that the individual is incapable of satisfactory dealing with it, then the effect on performance may be negative. Thus, extreme stress conditions are said to be detrimental to human health, but in moderation stress is normal and, in many cases, proves useful.

Hans Selye (1956), an endocrinologist, perceived stress to be a neural physiological phenomenon. More specifically he defined it as a general adaptive syndrome or non-specific response to demands placed upon the human body. These demands could either stimulate or threaten the individual.

The term has been further defined by Gold and Roth (1993) "a condition of disequilibria within the intellectual, emotional and physical state of the individual; it is generated by one's perceptions of a situation, which result in physical and emotional reactions. It can be either positive or negative, depending upon one's interpretation.

In modern usage, however stress has come to imply the subjection of a person to force or compulsion, especially mental pressure or by overwork, which leads to strain or mental fatigue. In medical parlance, 'stress' is defined as a perturbation of the body's homeostasis. This demand on mind-body occurs when it tries to cope up with incessant changes in life.

Stressors: The Activators of Stress

The stressors are events or situations in our environment that cause stress. Stress is said to stem out from negative events in our life. But also some positive events such as getting married or receiving an unexpected job promotion can also produce stress. Despite the wide range of stimuli that produces stress; it appears many events share several common characteristics such as:

- They are so intense that they produce a state of overload.
- They evoke incompatible tendencies in us, such as tendencies both to approach and to avoid some object or activity.
- They are uncontrollable—beyond our limits of control.

Stress is often associated with situations that an individual find difficult to handle. How the individual view things also affects their stress level. If they have high expectations; chances are that they may experience more than fair share of stress.

An individual's stress may be linked to external factors such as:

- The state of the world, the country, or any community to which one belongs.
- Unpredictable events.
- The environment in which the individual work.
- Work itself.
- Family.

Stress can also come from the individuals' own irresponsible behaviour, poor health habits, negative attitudes and feelings, unrealistic expectations and perfectionism.

External and Internal Stressors

Stressors can be classified as external and internal based on their source.

External stressors include adverse physical conditions (such as pain or hot or cold temperatures) or stressful psychological conditions (such as poor working conditions or abusive relationships).

Internal stressors can also be physical (infections, inflammation) or psychological (worry, helplessness). An example of an internal psychological stressor is intense worry about a harmful event that may or may not occur.

Acute or Chronic Stressors

Stressors can also be defined as short-term (acute) or long-term (chronic).

Acute stress is the reaction to an immediate threat, commonly known as fight or flight response. The threat can be any situation that is experienced, even subconsciously or falsely, as a danger. Common acute stressors include:

- Noise.
- Isolation.
- Hunger.
- Danger.
- Infection.
- Imagining a threat or remembering a dangerous event.

Under most circumstances, once the acute threat has passed, the response become inactivated and levels of stress hormones return to normal, a condition called the relaxation response.

Frequently, however, modern life poses ongoing stressful situations that are not short lived, and the urge to act (to fight or to flee) must be suppressed. Stress, then, becomes chronic.

Common chronic stressors include:

- Ongoing highly pressured work.
- Long-term relationship problems.
- Persistent financial worries.
- Loneliness.

Symptoms of Stress

i. Acute Stress

According to American Psychological Association, the acute stress results form "demands and pressures of the recent past and anticipated demands and pressures of the near future". During situations where an individual experience acute stress, they are likely to experience increases in heartbeat and breathing. In small doses, acute stress may feel exciting, but too much eventually becomes exhausting and taxing of the body, mind and spirit.

Common symptoms include:

- Emotional distress (irritability, resentment, anger and depression)
- Muscular problems (tension headache, back pain, jaw pain, etc.)
- Problems involving the stomach, gut and bowels (heartburn, diarrhea, constipation, irritable bowel syndrome).

ii. Episodic Acute Stress

If the individual endure acute stress frequently, then probably are experiencing episodic stress. They might feel disorderly, in perpetual crisis, chaotic, or out of control.

The symptoms are:

1. Being always rushing and late.
2. Feel over-aroused, short tempered, anxious, and tense most of the time.
3. Have "worry wart" tendencies (focus on negative possibilities and anticipate crisis on disaster in most situations).

Symptoms of stress when studied briefly can be classified as physical, emotional and relational symptoms.

i. Physical Symptoms

The physical problems outlined below may result from or be exacerbated by stress.

- Sleep disturbances
- Back, shoulder or neck pain.
- Muscle tension or migraine headaches.

- Fatigue
- Weight gain or loss, eating disorders.
- Skin problems (hives, eczema, tics, itching)
- Immune system suppression: more colds, flu and infections
- Upset or acid stomach, cramps, heartburn, irritable bowel syndrome, constipation, diarrhea.
- High blood pressure, irregular heartbeat, chest pain.

ii. Emotional Symptoms

Like physical signs, emotional symptoms can mask conditions other than stress. The following symptoms are uncomfortable and can affect the individual's performance at work or play, physical health, or relationship with others.

- Nervousness or anxiety, depression, moodiness.
- Irritability, frustration
- Memory problems
- Lack of concentration
- Trouble thinking clearly
- Phobias
- Overreactions.
- Feeling out of control.

iii. Relational Symptoms

The antisocial behaviour displayed in stressful situations can cause the rapid deterioration of relationship with family, friends, co-workers, or even strangers. A person under stress may manifest signs such as:

- Increased arguments.
- Isolation from social activities.
- Conflict with co-workers or employers.

- Frequent job change.
- Overreactions.

Health Consequences of Stress

The physical changes in response to stress can be an asset for raising levels of performance during critical events such as a sports activity, an important meeting or in situations of actual danger or crisis. If stress becomes persistent, however, all parts of the body (the brain, heart, lungs, vessels, and muscles) become chronically over- or under-activated. This may produce physical or psychological damage over time.

i. *Physical Effects*

Stress related conditions that are most likely to produce negative physical effects include:

- An accumulation of persistent stressful situations, particularly those that a person cannot easily control (for example, high-pressured work plus an unhappy relationship).
- Persistent stress following a severe acute response to a traumatic event (such as an accident).
- An inefficient or insufficient relaxation response.
- Acute stress in people with serious illness, such as heart disease.

ii. *Heart Diseases*

Stress can certainly influence the activity of the heart when it activates the sympathetic nervous system. Such actions and others could theoretically negatively affect the heart in several ways.

- Sudden stress increases the pumping action and rate of the heart while causing the arteries to constrict, thereby restricting blood flow to the heart. A 2002 study suggested that such actions may be responsible for some incidents of acute stress that have been associated with a higher risk for serious cardiac events, such as heart rhythm abnormalities and heart attacks, and even death, in people with heart disease.

- Emotional effects of stress alter the heart rhythms, which could pose a risk for serious arrhythmias in people with existing heart rhythm disturbances.
- Stress causes blood to become stickier, increasing the likelihood of an artery clogging blood clot.
- Stress appears to impair the clearance of fat molecules in the body, raising blood cholesterol levels, at least temporarily.
- These cholesterol deposits also serve as a deposit for Calcium, after which the plaques harden and become inflexible. These unhealthy arteries cannot carry sufficient blood to the heart muscle and become blocked.
- In women, chronic stress may reduce estrogen levels, which are important for cardiac health.
- Some evidence supports the association between the stress and hypertension. In some studies, people who regularly experienced sudden increases in blood pressure caused by mental stress may, over time, had developed injuries in the inner lining of their blood vessels.

iii. ***Susceptibility to Infections***

The immune system is the body's primary defense against disease-causing agents including viruses, bacteria, fungi and other parasites. Hans Selye described three phases of the body's reaction to stress. The first stage, alarm reaction, is characterized by the release of large quantities of cortisol and catecholamine, which result in cardiovascular changes. The second stage known as stage of resistance, involves counter-regulation of the immune response. Large quantities of glucocorticoids begin to act to suppress immunity. The third phase, the stage of exhaustion, occurs only when the period of stress is prolonged. This stage involves "stress related immuno-compromise," which is often followed by death.

Chronic stress appears to blunt the immune response and increase the risk for infections and may even impair a person's response to immunizations. A number of studies have proved that subjects under chronic stress have low white blood cell counts and are vulnerable to colds (Cohen et al., 1991). Many studies have also been conducted which connect chronic stress to immune system. According to Sapolsky (1998), long periods of stress inhibit many functions of the immune system, including the formation and subsequent release of lymphocytes, the period of time lymphocytes remain in circulation, and in the production of antibodies.

iv. *Gastrointestinal Problems*

The brain and intestine are strongly related and mediated by some hormones and nervous system. Stress disrupts the digestive system, irritating the large intestine and causing diarrhea, constipation, cramping, and bloating. Excessive production of digestive acids in the stomach may cause a painful burning.

v. *Peptic Ulcers*

It is well established that most peptic ulcers is either caused by the Helicobacter pylori or by the use of non-steroidal anti-inflammatory medications. According to Sapolsky, 1998, during stress, digestion is often inhibited and the body cuts down its acid secretion and as response mucus secretion is stopped. During recovery, then, the acid returns and the stomach defenses are down with mucus secretion being low. This coupled with a bacterial infection, could easily result in ulcer. During stress, blood is delivered to the muscles that are most necessary for function. Thus blood is diverted from the gut to other parts. Nevertheless studies still suggest that stress may predispose someone to ulcers or sustain existing ulcers. Up to 15 per cent of duodenal ulcers form in people who aren't infected with the bacteria. Also only 10 per cent of the people infected with bacteria get ulcers (Sapolsky, 1998). Therefore, stress is probably an important additional factor to ulcer formation.

vii. *Bowel Diseases*

Irritable bowel syndrome (spastic colon) is strongly related to stress. With this condition, the large intestine becomes irritated, and its muscular contractions are spastic, resulting in cramping and alternate periods of constipation and diarrhea. Although stress is not the main cause of inflammatory bowel disease (ulcerative colitis), there are reports of an association between stress and symptom flare-ups. It has been found that short-term stress did not significantly exacerbate ulcerative colitis symptoms; long perceived stress tripled the rate of flare-ups compared to patients who did not report stress.

viii. ***Eating Problems***

Stress can have varying effects on eating problems and weight.

ix. ***Weight Gain***

Often stress is related to weight gain and obesity. Many people develop cravings for salt, fat and sugar to counter their tension and, thus, they gain weight. The release of cortisol, a major stress hormone, appears to promote abdominal fat and may be the primary cause of weight gain.

x. ***Weight Loss***

Some people suffer a loss of appetite and lose weight. In rare cases, stress may trigger hyperactivity of the thyroid gland, stimulating appetite but causing the body to burn up calories at a faster than normal rate.

xi. ***Eating Disorders***

Eating disorders frequently co-occur with psychiatric disorders such as depression, substance abuse and anxiety disorders. Females are much more likely than males to develop an eating disorder. Chronically elevated levels of stress chemicals have been observed in patients with anorexia and bulimia. People with anorexia nervosa have intense fear of gaining weight or becoming fat, even though underweight. The process of eating becomes an obsession. Unusual eating habits develop, such as avoiding food and meals, picking out few foods and eating those in small quantities, or careful and frequent weighing. Bulimia Nervosa is characterized by eating an excessive amount of food within a discrete period of time and by sense of lack of control over eating. There is recurrence of inappropriate compensatory behavior in order to prevent weight gain, such as self-induced vomiting or misuse of laxatives, fasting or excessive exercise.

xii. Diabetes

Chronic stress has been associated with the development of insulin resistance, a condition in which the body is unable to use insulin effectively to regulate glucose (blood sugar). Insulin resistance is a primary factor in type 2 diabetes. It leads to increased levels of glucocorticoids, which increase the levels of glucose in blood. Obviously, these increased glucose levels raise the likelihood of damage to kidneys and blood vessels, as well as blindness, diabetic coma or even death. Stress can also exacerbate existing diabetes by impairing the patient's ability to manage the disease effectively.

xiii. Stroke

In some people, prolonged or frequent mental stress causes an exaggerated increase in blood pressures thus leading to stroke. One survey revealed that men who had more intense response to stressful situations, such as waiting in line or problems at work, were more likely to have strokes than those who did not report such distress.

xiv. Pain

Researchers are yet to find the accurate relationship between pain and emotion, but there are many factors that induce pain in stress.

xv. Muscular and Joint Pain

Chronic pain caused by arthritis and other conditions are intensified by stress. Psychological stress had been studied to play an important role in the severity of back pain. Some studies have also associated job dissatisfaction and depression to back problems.

xvi. Headaches

Tension-type headache episodes are highly associated with stress and stressful events. Among the wide range of migraine triggers the main factor found was emotional stress.

xvii. Sleep Disturbances

The tensions of unresolved stress frequently cause insomnia, generally keeping the stressed person awake or causing awakening in the middle of the night or early morning. In fact, it has been found that stress hormones can increase during sleep in anticipation of a specific waking time.

Other Disorders

i. Allergy

Research suggests that stress, may actually be a cause of the sick-building syndrome, which produces allergy-like symptoms, such as eczema, headaches, asthma and sinus problems, especially in office workers and students.

ii. Skin Disorders

Stress plays a role in exacerbating a number of skin conditions, including hives, psoriasis, acne, and eczema. Unexplained itching may also be caused by stress.

iii. Unexplained Hair Loss

Alopecia aerate is hair loss that occurs in localized patches. The cause is unknown, but some researchers had found that stress plays an important part. Hair loss is said to more often relate with intense stress such as mourning or any chronic job stress conditions.

iv. Teeth and Gums

Some researchers have found that stress being an increasing risk for periodontal disease, which is disease in the gums that can cause tooth loss.

Psychological Effects of Stress

Other than physical effects of stress on the body, there are psychological effects as well. Studies suggest that the inability to adapt to stress is associated with the onset of depression or anxiety. In one study, two-thirds of subjects who experienced stressful situation had nearly six times the risk of developing depression within a month of the stressful event. It had also been suggested with evidence that the repeated release of stress hormone produces hyperactivity in the hypothalamus-pituitary-adrenal axis and disrupts normal levels of serotonin, the nerve chemical that is critical of well-being. Thus stress diminishes quality of life by reducing feelings of pleasure and accomplishment and relationships are often threatened.

i. Depression

The same type of events that cause stress can also ultimately cause depression. It is healthy and natural to grieve and be depressed for a period of time, but it is when that period of time is extended that it becomes unhealthy (Wagemaker, 1994). Out of all the minor cases of depression, 5-20 per cent will endure a more serious and dangerous form of the disease (Sapolsky, 1998).

When a stress response is experienced, there is an increased production of epinephrine and cortisol by the adrenal glands. Normally the levels of these hormones will stay high as long as the stressful stimulus is present, but when those levels of hormones remain high for an extended time, the body 'burns out' (Zuess, 1998). The result of this high is a wide mood swing (Sturgeon, 1979). Frequent mood swings are common symptoms of depressives.

Wina Sturgeon, 1979, explains how stressors and chemical imbalances are related through categorizing depression. The four categories of depression are: reactive, endogenous, toxic and psychotic depression.

Reactive depression is the immediate response to some stressful event that causes a personal setback. Trouble falling asleep, loss of appetite, sadness and anxiety are some of the characteristic symptoms of this depression. Majority of depression cases fall in this category.

Endogenous refers to the depression as "Stemming from within" (1998). The only apparent cause is a chemical imbalance within the brain. Dry mouth, constipation, abdominal pains, headache, loss of memory, concentration and feeling of sadness and anxiety are some of the physical and psychological symptoms of this type of depression. Loss of weight, interest in sex, energy and the inability to make decisions are also commonly experienced.

Toxic depression is caused by the use of drugs. Alcohol, narcotics and even prescription medications can cause such depression. In some cases the person will experience hallucinations and mood swings.

The stigma that is associated with depression is mostly credited to psychotic depression. People with psychotic disorders are at times out of touch with reality caused by brain disorders or over-exhaustion. This is the source of the majority of nervous breakdowns. This state can be well described by drastic mood swings, hallucinations, and compulsive behaviours.

ii. Memory, Concentration and Learning

Stress has significant effects on the brain, particularly on memory and concentration thereby learning process is affected.

iii. Effect of Acute Stress on Memory and Concentration

Studies indicate that the immediate effect of acute stress impairs short-term memory, particularly verbal memory. A 2002 study of healthy women reported that high levels of natural cortisol were associated with enhanced declarative memory, which is the storage of facts and connections between facts and stimuli.

On the positive side of stress, it is said that high stress hormone levels are associated with a greater concentration on sensory stimuli.

iv. Effects of Chronic Stress on Memory

The typical victim of severe chronic stress often suffers loss of concentration at work and at home and may become inefficient and accident-prone. In children, the physiologic response to stress can clearly inhibit learning. Although some memory loss occurs during aging process, chronic stress plays an important role. Studies have associated prolonged exposure to cortisol to shrinkage in the hippocampus, the center of memory.

ATTITUDE

Attitude is more of less generalized tendency to think or act in a certain way to some object or situation often attended by feelings. It can be defined as a feeling on evaluative long lasting reaction to a particular object of people. Allport defined attitude as a mental and neural state of readiness, organized through experience exerting directive or dynamic influence upon the individual's response to all objects and situation with which it is related. It is identified with prejudice, biases, and states of readiness, beliefs or ideas with an emotional being.

Social Psychologists call attitudes; generally involve an emotional or affective component (for instance, liking or disliking), a cognitive component (beliefs), and a behavioral component (tendencies to act toward these items in various ways). More simply, attitudes can be defined as lasting evaluations of virtually any and every aspect of the social world – issues, ideas, persons, and social groups, objects (Fazio and Roskos Ewoldsen, 1994; Tessar and Martin, 1996).

An attitude is always a stand or position which an individual takes towards a person or an issue. Accordingly, one has either a positive or negative attitude towards some aspect of one's environment. Finally an attitude is always a pattern of ideas, motives and perceptions.

The major functions of attitude are adjustment, ego-defensive, expressing one values and our knowledge about some object or person.

Attitude Formation

Majority of attitudes held by a person are acquired from the members of the family and from the peer group in early childhood and later. It occurs in several processes.

i. Classical Conditioning Approach

Classical conditioning approach may be especially influential in shaping the emotional or affective aspect of attitudes (Cacioppo, Priester and Bernston, 1993; Krosnick, et al., 1992). On successive occasions, a neutral stimulus is paired with an unconditional stimulus. Over time, the previously neutral stimulus may begin to elicit response similar to that produced by the unconditional stimulus. For example, a young child sees her mother show sign of emotional discomfort, through facial expressions, when she encounters members of minority group. Though, initially the child has no emotional reaction, it acquires from her mother's repeated action and thus results in a racial prejudice.

Objects, peoples, or events associated with pleasant experience may take on favourable evaluations, while those associated with unpleasant experiences may be evaluated negatively. Griffitt (1970) had people interact in small groups in either a comfortable room or one, which was hot and uncomfortable. When asked to rate how much they liked the other people present in the room, individuals in the hot room reported liking the others less than did individuals in the comfortable room.

ii. ***Operant Conditioning Approach***

Instrumental conditioning, in which the reward consequences of any behaviour shape its subsequent enactment, obviously influences attitude formation. By rewarding children with smile, approval and compliments for stating some right views or doing right things, attitudes can be incorporated in them. Membership and acceptance in a particular group is often contingent upon the attitudes one expresses. Peer groups differentially reinforce the expression of certain attitudes relevant to the group and the society.

iii. ***Observational Learning Approach***

Attitudes are also obtained from Modeling. The individual easily tends to adopt the views and preferences expressed by people they like or respect, because they are exposed to those views quite often and want to be like those persons. They accept themselves as the role model and their attitudes and behaviours are easily incorporated in them. Social comparison also plays an important role in attitude formation. It is a basic tendency to compare with others in order to determine whether our view of social reality is correct or not.

iv. ***Genetic Influence***

Evidence suggests that genetic factors too play an important role in shaping attitudes. The attitudes of identical twins, who share the same genes, have been found to correlate more highly than those of non-identical twins (Crealia and Tesser, 1996). Further this is having been proved even when the twins were separated early in life and raised in sharply contrasting environments from then on (Hershberger, Lichtenstein, and Knox, 1994).

Determinants of Attitude

The two determinants of attitude formation and attitude change are psychological and cultural.

i. Psychological Determinants

Psychological determinants include factors such as motivation, emotion, need, thinking, dominance and submission, which play a part in originating or changing a person's attitude.

A number of studies have demonstrated that personality traits like introversion and extroversion, dominance and submission play a role in attitude formation. For example, politically radical individuals tend to be more submissive and introverted and to possess more inferiority feeling than conservatives.

However, correlations of personality and attitudes can be misleading, since they are subject to distortion and misinterpretation. Thus it may be mistakenly inferred that all introverted and submissive people have radical ideas, and conversely, that all the ascendant and extroverted individuals are conservative and reactionary.

ii. Cultural Determinants

Specific cultural determinants include variables such as social status, family environment and education. Students from wealthy background have been found to be more prejudiced and more authoritarian than those of middle class origin. A study made by Social Research Council in 1947 revealed that less educated persons more often advocated the use of atom bombs to compel other nations to submit to our country's wishes than the educated. These examples indicate that cultural factors are significant determinant of human attitudes. The force of cultural norms is often effective and is rarely disregarded by people in a group. Not only the origin, but also the persistence, of attitudes arises from cultural values and norms.

Attitudes are always the result of the combined and interactive effect of both personality and cultural variables. We hold our beliefs or attitudes because we are conditioned to do so by our family, class membership, education, culture or religion. Though we are conditioned, we unconsciously adhere to only those attitudes, which are congruent with our perception of what is "right" or "true", which fit into the cognitive structure or preconceptions. Psychological and cultural variables always interact in producing, retaining or changing attitudes.

Attitude Change

Attitude formation and change occur in the context of existing interpersonal relationships, group memberships, and particular situations; and they span various time periods. Sometimes the amount of attitude change is extreme. Whether extreme or moderate, fast or slow, it is possible to identify the basic units involved in attitude change processes. Persuasion is the process through which one or more persons attempt to alter the attitudes of one or more others. Persuasion depends on the content of the message or information, its source and the person who is being persuaded.

One of the sources of change is obtaining new information from other people or media. Attractive sources are more effective in changing attitudes than unattractive ones. (Kiesler and Kiesler, 1969). Change can be enhanced by messages that arouse strong emotions, especially fear. Speeches delivered by experts to people have more effect on change in their attitude (Hovland and Weiss, 1951). Individuals with low self-esteem are easily persuaded than those with high self-esteem. People who speaks rapidly are often more persuasive than those who speak more slowly (Miller et al., 1976).

TEACHING AS A PROFESSION

i. *Profession*

A profession is defined as something more than the distinction between volunteer and paid service providers. A profession is a specialized type of occupation, which maintains formal standards for admission, which requires training, which maintains codes of ethics and formally takes away membership from those who are proved unworthy. Professions involve intellectual operations, derive raw materials from learning, work up the material to the practical and definite end, possess an educationally communicable technique, move toward self-organization, and becoming altruistic in nature (Stinnett, 1962).

The status of the profession is built by the work of the members of the profession. If we admit the right type of person to the profession, then naturally the public opinion on that profession would be favorable. A professional attitude is a specialized occupational tendency to act, arising from skill in a highly complex and limited field. Members with favorable professional attitude are an asset while those with unfavorable attitudes bring down others opinion.

ii. Teaching

Teaching is a profession that plays an important role in a country's development. Teaching involves the transmission of knowledge from one generation to another. The teacher is the living ideal and a potential guide to provide directive growth and development in a student of today as worthy citizens of country. Teachers are the people that affect children's future, and in doing that they also affect the countries economy (DuFour, 2000). They build on student's prior knowledge, life experience and interests to achieve their learning goals. They actively engage all students in problem-solving and critical thinking within and across subject matter areas. It is the responsibility of the teacher to guide and inspire the student to enrich his discipline and to indicate values in accordance with our cultural heritage and our social objectives.

Teachers facilitate challenging learning experiences for all students in environments that promote autonomy, interaction and choice. The teacher is a stimulus and a model for development for the students (Cohn and Kotthamp, 29). Teachers nurture the intellectual, physical, emotional, social, and civic potential of each student. They create, support, and maintain challenging learning environments for all.

The National Educational Policy (1986) has said, "The status of the teacher reflects the socio-cultural ethos of a society. It is said that no people can rise above the level of its teachers. The Government and the community should endeavor to create conditions, which will help motivate and inspire teachers on constructive and creative lines. Teachers should have the freedom to innovate, to device appropriate methods of communication and activities relevant to the needs and capabilities of and the concerns of the community".

ROLE AND RESPONSIBILITY OF TEACHERS

The teacher has an important role to play in the total programme of national development and social change. He should go up to the institution with a sense of values and purpose and be fully equipped to play his role and not only as a professionally trained person but as an enlightened and dedicated member of the society. The first and foremost responsibility is in relation to his students. The primary objective should be to treat each individual student has an end in himself and to give him a widest opportunity to develop his skills, abilities and potentialities to the full extent.

The teacher has to influence the life and character of students, and equip them with ideas and values, which will fit them to be worthy citizens. The teacher's role is to be an educational sociologist according to Kleinsasser (1993). They have to educate them on the need to recognize the equality of men and women in our democracy, to discard all caste exclusiveness and pride, untouchability, and communal distinctions and antagonisms.

Hughes (1986) states that a teacher should be a philosopher, geographer, historian, philologist and literary critic. To Altman (1981), the teacher functions as a "skillful developer of communicative competence in the classroom's dialectologist', 'value clarifier", and "communication analyst" (pp. 11-13).

Teachers should assist students to become self-directed learners who are able to demonstrate, articulate and evaluate what they learn. It is teacher's full responsibility to make the class interesting, to view the teacher as a facilitator to knowledge, a resource in subject. The teacher should be very facilitative to the student's reach that he or she is able to reveal his or her emotional problems or other problems. The teacher's patience in their duties attacks ignorance and misunderstanding.

The teacher should also accept his responsibility in the realization of our social objectives, which implies that education should be related to the life needs and aspirations of the people. From this point of view, it becomes important for a teacher to have an active part in (i) Programmes of community development, (ii) Adult education and extension, (iii) Social and National services, (iv) Co-curricular and extra-curricular activities, (v) Programmes of non-formal education and (vi) Social and national integration.

Recently, the role of teacher has to change, still in accordance with the educational and scientific development. Teacher's knowledge, skills, and practices develop throughout their professional careers. The nature of teaching requires continuous growth in order to engage and challenge increasingly diverse students in a rapidly changing world. The teacher must grow in the "what" and "how" of teaching because of explosion of knowledge is adding immensely to "what" and vast changes in education, media, its tools and conditions around brought a big change in "how".

Teachers in modern schools, whether primary or secondary, are often amazed at children's eagerness to learn, their creative problem-solving, persistent efforts in self-development and their academic, artistic and athletic accomplishments. Trained teachers who are curious and scientifically oriented often wonder why and how children learn as much and even more effectively. The teachers can obtain positive results, if they can apply psychological principles in the classroom, laboratory, on the playground or in any learning situation.

Basis of Effective Teaching

The effective teaching comprises more than the assigning and hearing of lessons in a relatively recent educational recognition. Traditionally, the mastery of appropriate subject matter was the primary, sometimes the sole, requirement for teaching. But although teacher and learner relationships were stressed, there always have been great teachers who influenced not only thinking and behaviour of the pupils but also the attitudes of their societal group.

The number of mastery teachers was small, however, until analytical studies of teacher competencies were undertaken. The findings of already completed and still-continuing investigations in this field have resulted in establishment of teacher-training divisions in colleges and universities. Educators responsible for teacher training constantly are attempting to improve their methods of selecting candidates, their curricular offerings, and their own instructional approaches.

The characteristic of outstanding teachers was given in a session of Michigan Department of Education (2001). They are:

- Caring.
- Unending Energy.
- Dedicated.
- Enthusiastic.
- Knowledgeable about content.
- Knowledgeable of current trends in education and educational theories.
- Commitment to student and self-growth.
- Understanding of personal needs of each student.
- Time management and organizational skills.
- Innovative.
- Able to reach out and build relationships with parents.
- Quick decision-making ability.

The many adjective aspects of living in our modern complex society require that teachers be well qualified to assume their teaching responsibilities. Programs of teacher education both pre-service and on service have become increasingly effective.

In order to insure effective teaching, consideration is given to certain phases of teaching as their prospectus. Teaching is an integrative and integrated process in which the functioning of one phase or facet influences the functioning of others.

Mastery of Subject Matter

The term mastery implies that the teacher not only knows what he is teaching but also can organize and adapt materials in accordance with his student's ability and interest level. The teacher, as an expert in his field, should also be able to place himself in the position of a beginner.

Physical and Health Status

Delicate health or poor physical stamina may incapacitate a person for the adequate performance of teaching activities. Mental strain also can affect their attitude towards their profession and students. Thus an effective teacher needs to be free from physical, mental or any health difficulties. A teacher's absence may interfere with teaching-learning process.

Personal Attributes and Emotional Control

A teacher's behaviour and expressed attitudes exert a powerful attitude upon young people. Hence, it is very important that a teacher to be an honest, sincere, disciplined person whose ethical standards are worthy of imitation. Careful grooming, pleasing voice, good diction and habitual use of acceptable grammar also are important. During the course of teaching, a teacher can be subjected to challenge of dealing with uncooperative, uncontrollable pupil behavior. The teacher needs to be with self-control and exercise patience in helping them overcome personal or learning difficulties.

Understanding of Human Nature and Development

A man or woman may be temporarily fitted to teach and very much interested in working with young people, but unless he/she has adequate knowledge of the sequential pattern of human development he/she is likely to be an ineffective teacher. The causal factors underlying the many vagaries displayed in child and a teacher must understand adolescent behaviour so that he can provide opportunities for his pupils to achieve greater reactional consistency.

Knowledge and Application of Learning Principles

It is sufficient to re-emphasize the fact that discretion is needed in the application of principles or theories of learning. What should be taught, and the why, when and how of teaching are dependent on factors such as individual and societal needs, learner readiness, and available teaching-learning opportunities and materials. The knowledge acquired should be applied in such a way that it is reached in a proper and directional way and the purpose has been achieved. So applying the learned theories, learning principles on the teaching field is important.

Sensitivity to and Appreciation of Differences

Regardless of his own ethnic and cultural background and his religious affiliation, the teacher's attitude towards his pupils, fellow teachers, and other school personnel must be unbiased and objective. Nor should he attempt to influence the thinking of young people in accordance with his personal views on social and political issues.

Continued Professional and Cultural Improvement

A teacher's education is not completed when he is certified to teach. What may have been adequate teaching methods at that time may fall to meet later educational demands or may need to be changed in the light of the results of continued psychological research.

SPECIAL EDUCATION

Special education, describes an educational alternative that focuses on the teaching of students with academic, behavioral, health, or physical needs that cannot sufficiently be met using traditional educational programs or techniques. Special education teachers design and modify instruction to meet a student's special needs. They also work with students who have other special instructional needs, including those who are gifted and talented.

The various types of disabilities that are categorized under special education programs include specific learning disabilities, speech or language impairments, mental retardation, emotional disturbance, multiple disabilities, hearing impairments, orthopedic impairments, visual impairments, autism, combined deafness and blindness, traumatic brain injury and other health impairments. Students are classified under one of the categories, and special education teachers are prepared to work with specific groups.

RESPONSIBILITY OF SPECIAL EDUCATORS

Special education teachers are involved in a student's behavioral as well as academic development. They help the students develop emotionally, be comfortable in social situations, and be aware of socially acceptable behaviour. The mentally and physically challenged students need to be trained to sustain their day-to-day life activities, and socializing activities.

Special educators have many roles and responsibilities in any given school setting as they work with children of different kinds and severity of disabilities. Their responsibilities can be usually categorized into three divisions: Direct teaching, paper work, and collaborating with other professionals and parents.

Direct Teaching

The primary responsibility of every teacher being the teaching, the special educators have a special task of developing the instruction materials to match the learning styles, capabilities, and special needs of each of their students. They spend close to the majority of their classroom time actually teaching their students (Allinder, 1994). These time spent does not include the other assignments or meeting with other professionals.

In general education, it is the school system dictates the curriculum, but in special education, the child's needs dictate the curriculum (Liberman, 1985). Thus, the special educator has to alter the curriculum separately for each children based on their abilities, age, setting and many other variables. Early identification of a child with special needs is an important part of a special education teacher's job. Early intervention is essential in educating children with disabilities. Depending upon the kind and the severity of the disability, teaching aids and methods are facilitated.

Special education teachers use various techniques to promote learning. Depending upon the disability, teaching methods can include individualized instruction, problem-solving assignments and small-group work. They develop an Individualized Education Program (IEP) for each special education student. The IEP sets personalized goals for each student and is provided to the student's individual needs and ability. When appropriate, the program includes a transition plan outlining specific steps to prepare students with disabilities for middle school or high school or jobs if capable.

Special education teachers design and teach appropriate curricula, assign work based upon each student's needs and abilities, and grade papers, and homework assignments. They review the IEP with the student's parents, school administrators, and the student's general education teacher. They also should assess each child's gradual academic achievement and then modify or design the instruction materials, in accordance with, so that the child can achieve the expected outcome.

Paper Work

Special education teachers deal with more amount of paper work than that of normal teachers. They have the same work demands that of the general education teachers; attendance reports, grading the students through assignments and tests, daily notes of lesson to be taught, discipline reports etc. But they also have additional work of preparing individualized Education Program (IEP) for each student. They also should maintain accurate and complete records of each student that document their progress towards the goal and objectives specified in the IEP.

Collaboration with Other Professionals

A large part of special education teacher's job involves interacting with others. Special educators never work alone on the academic and social development of the children. They have established teams to help them plan their instruction materials and the educational intervention for students. They work with other professionals depending on the child's disability and the school settings. They need to work with speech therapists, school psychologists, occupational therapists, school social workers, general education teachers, and community workers, to plan and implement the IEP for the children.

Parents having a great expertise on their child, play an important and powerful part in the whole team of the child's consultation program. The special educators and the parents should work together, and contribute to the educational planning for children with disabilities. Special educators should get report from each child's parent and also follow-up the report both in the academic, other physical improvement from other professionals and decide a child's overall achievement. Many special education teachers who leave the profession say that one of the reasons they did so was the lack of time to meet all of their responsibilities (Billingsley et al.).

Working Conditions

Special education teachers work in a variety of settings. Some have their own classrooms and teach only special education students; others work as special education resource teachers and offer individualized help to students in general education classrooms; still others teach together with general education teachers in classes composed of both general and special education students. Considerably fewer special education teachers work in residential facilities or tutor students in homebound or hospital environments.

A large part of special education teacher's job involves interacting with others. They communicate frequently with parents, social workers, school psychologists, occupational and physical therapists, school administrators and other teachers. As schools have become more inclusive, special education teachers and general education teachers increasingly work together in general education classrooms.

Preparing special education students for daily life after graduation also is an important aspect for job. Teachers provide students and their parents with career counseling or help them learn routine skills, such as balancing a check book. Teachers work closely with parents to inform them of their child's progress and suggest techniques to promote learning at home.

Thus special education teachers enjoy the challenge of working with students and the opportunity to establish meaningful relationships. However, the work can also be emotionally and physically draining. Special education teachers are under considerable stress due to heavy workloads and tedious administrative tasks. They work under the threat of litigation against school or by the student's parents whether correct procedures are followed or not.

Some special educators feel they are not adequately supported by school administration, and feel isolated from general education teachers. The physical and emotional demands of the job also cause some special education teachers to leave the occupation.

STRESS IN TEACHING

A survey was conducted with 664 public school teachers and 250 private school teachers to see if they had a sense of calling for the career. Eighty-six per cent (86%) of these teachers said that those with a true sense of calling should pursue the career, and feel that this is the case with the new teachers today (Wadsworth, 2001). School administrators were also asked this question. The reaction they gave about the new teachers was very upbeat. They said the teachers were very motivated, had a personal and loving drive for their job, and were trained very well. The administrators were very pleased with the new generation of gifted teachers.

Another study asked teachers to rate the most important attributes of being in that career. Eighty-three per cent (83%) say that being able to do work that they love is essential. Eighty-one per cent (81%) report that having time with family is also important. Then seventy-two per cent (72%) say that being able to help society is an important thing. Other attributes like the importance of salary and opportunities for advancement were not even the least bit important (Wadsworth, 2001). Ninety-eight per cent (98%) of administrators see that the new teachers are incredibly motivated, caring, and extremely energetic. For these people teaching is a perfect career and they are there for the right reasons. Teachers receive very good hours and vacation time during the year so they can spend time with family.

However, the status of teacher deteriorated over the last two decades mainly due to poor service conditions of teachers, phenomenal expansion of educational system, negligence of duty by many teachers and change in the value system of the society. In a research conducted using 1999-2000 Schools and Staffing Survey data file and the Teacher Follow-up Survey, according to principal reports, 19 per cent (19%) of Catholic school teachers, 23 per cent (23%) of other religious school teachers, and 21 per cent (21%) of nonsectarian school teachers changed schools or left the teaching profession between the 1999-2000 and 2000-01 school years.

Private school teachers who were reported to have left their schools (movers and leavers) were more likely than stayers to report relatively low levels of administrative support, satisfaction with salary, student discipline, control over classroom policies, and input in school policies (Daniel McGrath and Daniel Princiotta, June 2005).

It is therefore necessary to make an intensive and continuous effort to raise the economic, social and professional status of teachers in order to attract young men and women of ability to the profession and to retain them in it as dedicated, enthusiastic, and contented workers.

Turning specifically to stress in teaching, Kyriacou, 2001, defined it as the experience by teachers of unpleasant, negative emotions, such as anger, anxiety, tension, frustration, depression, resulting from some aspect of their work as a teacher (p. 28).

Dunham (1984), proposed three ways of defining stress. Each model has different implications of teachers and educational managers.

The Medical Model

The medical model focuses on physiological and psychological responses, which can arise as a consequence of stress. A plethora of symptoms, such as depression, tension, insomnia, loss of appetite, and weight loss, are essential components of the definition.

The Interactive Model

This model perceives stress as interactive and situational. It recognizes that on the one hand teaching as a profession and some schools in particular may exert pressures on teachers; while on the other, individual teachers react in different ways and bring a variety of adaptive resources to help them cope with those pressures.

The Professional Model

Of the three models, the third approach is perhaps the most helpful. It implies that responsibility for the maintenance of acceptable levels of stress in teaching is a two-way process. Employers have a statutory duty to ensure that the working environment in schools does not adversely affect employee's health; but teachers must also apply their adaptive resources to help them cope with the inherent pressures of their chosen profession.

CAUSAL FACTORS OF TEACHER'S STRESS

There are unquestionably a number of causal factors for teacher stress. The most important factor is job satisfaction, good attitude towards the teaching profession. These factors lead to a healthy professional and personal life. When their job satisfaction gets lessened, their attitude towards their profession decreases and causes stress in their profession and thus affecting their physical, health and personal aspects.

Review of research has revealed many factors that cause stress in teachers toward their profession, both in their personal and professional life. They are listed as:

- Administration
- Working conditions
- Monetary aspects
- Status of the profession in the society
- Teaching workload
- Pupil's attitude and behaviour
- Personal problems

Administration

The teachers or any other professionals to continue in the profession need to be supported by their administration. They must be provided with adequate teaching materials, sufficient school buildings and equipments, sufficient time and work. The school organization as a causal factor includes large student-teacher ratio, the size of the class, frequent changes in the use of buildings, and lack of communication and consultation in the school. Some teachers often report that they also lack in time for meeting the parents and other professionals if required. In a study conducted by Armes, 1985, 60 per cent of the respondents surveyed found their Head teachers to be the source of stress. Morgan (1977) reported that teachers felt that the support of both their colleagues and of their Head was very important to them.

Rudd and Wiseman (1962), discovered differences in degrees of satisfaction and dissatisfaction between infant school teachers, junior, and secondary modern school teachers and between men and women. They analyzed responses of 416 full-time teachers and produced a list of significant responses in rank order of frequency of reporting by these teachers. Among them, the main factors reported were teacher's salaries, administration, teaching load, and lack of teaching aids, inadequate school buildings and equipments, etc.

Poor Working Conditions

Poor working conditions, generally in terms of relations with colleagues constitute one of the major sources of occupational stress. Dunham (1976), Cox (1977) Kyriakou and Sutcliffe (1979) have reported in their individual research, that poor working conditions are one of the significant source of stress. Poor human relation among staff, lack of communication, both quantitatively and qualitatively, or the difficulties experienced in achieving effective communication, plays an important causal factor of stress.

Monetary Aspect and Status

The salaries paid to the teachers, play an important role for women teacher's stress, if they are the primary bread winners in their family, and the status of the profession in the society also forms a causal factor. Some teachers felt they are made scapegoats for the societal and educational problems in the country. There being many conflicts between the society and the school arising, lack of support of concern from parents and other professionals, changing social demands, etc., also form some of the causes of stress. Farkas, Johnson and Foleno (2000), found that 66 per cent of the population of thousand teachers, they surveyed maintained that they did not feel respected or appreciated and 76 per cent argued that the teachers are made scapegoats for all the problems in education. Henke, Chen, Geis, and Knepper (2000), found that only 14.6 per cent of 11,000 teachers surveyed, were satisfied with the esteem in which society held the teaching profession.

Work Load

Dewe (1986), found that workload consistently came as the top as the most frequent problem, the most anxiety inducing problem in the study of 800 teachers in New Zealand. The volume of paper work, the hours of work, the extra-curricular activities, sometimes more administration work, are reported to be one of the causal factor of teacher's stress. The teachers work load include, paper work, preparing of instruction materials, assessment of the students, evaluation and reporting work. Reporting itself takes a huge amount of time, it includes reporting to the principals, parents about their children, and to other professionals in case of special educators. These all consume more amount of energy and time. Cooper and Kelley, 1993, in their study on occupational stress amongst teachers in United Kingdom, found that the main source of stress being workload and handling relationships with staff.

Pupil's Behaviour

The pupil's behavior and attitude towards their education also plays an important role as a causal factor of stress. The teachers suffered daily anxiety about the children they handle, their disruptive behavior, and violence behavior. Though it is not a main causal factor it had significant correlation with poorer health. The pupil's poor attitude toward their work, poor motivation, pupil's misbehavior, leads to spending more of time to individual pupils and also clustering the non co-operative and aggressive children as a separate group and thus spending more time and effort, individually creates more stress (Kyriakou and Sutcliffe, 1978).

Personal Problems

Personal finances and perceived opportunities are the main influential factors that decide a teacher's attrition or retention. Personal adjustments the teachers undergo in the society, family etc., also plays a causal factor of stress in teachers. The health problems due to heavy workload, the compulsions of employment, financial crisis, problems arising due to employment in the family, are some of the personal problems they face thus leading to mental distress. Westling and Whitten, 1996, found that special educators who were "primary bread winners" were more likely to stay than those who were not.

Due to long working hours and more workload, the teachers mostly women, pose a main problem of facing the family problems, family or person move, rearing children, health, maintaining relationship with other family members, other financial crisis, etc, which leads to some health problems and also some familial problems such as marital breakup or personal relationship (Troman, 1998).

In a report given by Tom Cox of the University of Nottingham and Jack Dinham of the University of Bath, titled, "The Management of Stress in Schools" they list the factors, which are conducive to occupational stress in teachers as four.

- Comprehensive reorganization
- Changes in curricular and in teaching methods
- Unsuitable working environment
- Social and cultural changes and the impact of these changes upon education

STRESS IN SPECIAL EDUCATION

Special educators are found to be affected with stress more and frequently, due to some of the characteristics of their work, such as paper work, dealing children with different disabilities, frequent reporting to the parents and administrators, working with other professionals, etc. Some of the main sources of stress in special educators are excessive paper work, attitudes and behavior of other general teachers and administrators towards them, dealing with emotional and behaviour disorder children, etc.

Pullis (1992), conducted a study on 244 teachers of behaviorally disordered students to determine how the occupational stress affected their lives. Thus identifying the major sources of stress in special educators as:

- Excessive paper work
- Dealing with parents
- Behaviour problems of children
- Lack of recognition for good teaching
- Poor career opportunities
- Evaluations by administrators and supervisors

Thus the causal factor of teacher attrition or retention rate in terms of stress as a whole can be broadly factorized as three levels:

- Individual—teacher factors—Professional qualification and personal characteristics.
- School factors—peer collegiality and administrative support.
- System factors—Professional development and salary

EFFECTS OF STRESS IN TEACHING

Stress has been identified as one of the factors related to teacher attrition and is believed to be a cause of high teacher turnover and absenteeism. In a study conducted in Pacific region, by the National Center for Education statistics, 1998, it was found that one out of every five full time teachers leaves the teaching profession to pursue a career outside education field, the main reason being stress.

Stress may lead to problems in workplace and personal health problems. Some of the problems faced in workplace are poor morale, job dissatisfaction, less attitude, absenteeism, lowered productivity, and thus health problem resulting in high medical care costs (Kedjidjian, 1995). Absenteeism and other problems in turn lead to increase in higher percentage of poor outcomes in students (Madden, Flanigan, Richardson, 1991). Travers and Cooper, 1989, found that 23 per cent of their sample of 1800 teachers reported significant illness due to stress.

Stress in teaching also affects health thus leading to physical and emotional problems in teachers. Some of the main stress related physical and psychological problems in teachers are

- Cardiovascular diseases
- Spinal problems
- Musculoskeletal conditions
- Psychological distress
- Mood and sleep disturbance

The effects of stress in other than the personal sense are difficult to estimate. In teaching, the loss is defined in terms of departure of skilled teachers, impairment of teaching skills or sometimes even premature death. The effectiveness of teaching is said to decrease when stress increases (Firth – Cozens, 1992).

The major immediate effects of stress were also identified as:

- Feeling exhausted
- Feeling frustrated
- Feeling overwhelmed
- Carrying stress home
- Feeling irritable
- Feeling guilty about not doing enough

The objective of teachers should be making learners grow and develop mentally, socially and emotionally making them enjoy the thrill and advantage of learning. The teachers thus must be stress free and with a positive attitude towards their profession to held their responsibility in producing best and adequate students. They must be trained and given stress management programs to overcome the difficulties in teaching.

Thus, very few persons with adequate attitude and ability can be attracted to the teaching profession by motive, social service and love for teaching, overcoming the stress factors. So the provision of adequate remuneration, opportunities for profession advancement and favorable conditions of service and work, are the major programs, which will help to initiate and maintain this feedback process.

We have reached the threshold of the development of new technologies, which are likely to revolutionize classroom teaching. Unless capable and committed teachers are in service, the education system cannot utilize them for becoming a suitable and potential instrument of national development.

ATTITUDE ON TEACHING

Teacher's attitude towards teaching profession should be good as to perform their responsibilities. Over the course of last two years, Ellen Moir, Director of Santa Cruz Consortium New Teacher Project has taken a report from new teachers to find their attitude toward teaching and it has been divided into four different phases.

In the first phase of teaching, they used to romanticize the role of the teacher and the position. New teachers enter with a tremendous commitment to make a difference and a somewhat idealistic view of how to accomplish their goals.

In the second phase of teaching, the survival phase, they would learn a lot at rapid place. During this phase, most new teachers are struggling to keep their heads above water. They become very focused and consumed with the day-to-day routine of teaching. Although tired and surprised by the amount of work, first year teachers usually maintain a tremendous amount of energy and commitment during survival phase.

In the Disillusionment phase, the third phase, they realize that the things are probably not going as smoothly as they want and low morale contributes to this period of disenchantment. They face back-school night, parent conferences, and their first formal evaluation by the administrator. They express self-doubt, have lower self-esteem and question their professional commitment. This phase may be toughest challenge they face as a new teacher.

In the rejuvenation phase, the fourth phase, there is rise in the new teacher's attitude towards teaching. It is a time for them to send through materials that have accumulated and prepare new ones. A better understanding of this system, an acceptance of the realities of teaching, and a sense of accomplishment help to rejuvenate new teachers. Through the experience in their first half of the year, beginning teachers gain new coping strategies and skills to prevent, reduce, or manage many problems they are likely to encounter in the second half of the year. Many feel a great sense of relief that they have made it through the first half of the year. During this phase, new teachers focus on curriculum development, long-term planning and teaching strategies.

FACTORS INFLUENCING ATTITUDE

There are many factors that influence attitude towards a profession. The most important factor is job satisfaction, which is co-related with attitude. Good attitude or job satisfaction on teaching is actually performance of job with the sort of willingness and contended mind on the part of teachers. When there is more job satisfaction, and good attitude towards the co-workers and the school, there will be a positive attitude towards the profession. Review of research has revealed that there are some other factors other than job satisfaction as factors of attitude. They are listed as:

- Administration
- Working conditions
- Monetary benefits
- Occupation level
- The work group
- Age
- Race and Sex
- Educational Level
- Personal Adjustments

Administration

The first and foremost factor is the style of leadership in the administration. Generally employee centered leadership enhances a great amount of job satisfaction and attitude towards their profession. Their skills should be utilized, and not suppressed by the authority. They should be rewarded as per their achievements by incentives and promotions. Promotion provides economic and moral development in teachers. Their view on the school as an autocratic or a democratic, their decisions being considered, staff conferences being useful also play an important factor for the formation of good or a bad attitude towards their profession.

Working Conditions

Working conditions play an important role in the attitude towards teaching. The working conditions include facilities provided by the school, the condition of the students over there, the subordinates and co-teachers working with them, the teaching aids, the timetable they get, the remunerative activities they are allowed to work with and the politics that prefer among the teachers working.

Monetary Benefits

Monetary benefits also play an important role in attitude towards a job. In the developing world, monetary aspects being important, plays an important part in life that increases the job-satisfaction and attitude. Promotions and incentives increase the attitude towards teaching profession.

Occupation Level

Ample researches suggest that people in higher-level job experiences the highest levels of job satisfaction and attitude. One significant reason for this is that high-level jobs carry most prestige and self-esteem will be enhanced to the extent that other people view our work as important.

The Work Group

The work group also plays an important role in attitude towards job. The attitude and the performance of the subordinates, and the superiors working with teachers have an effect on the attitude towards their profession. The politics that prevail among the staffs plays an important role in job satisfaction.

Age

The relationship between age and their job attitude is both complex and fascinating. Research reveals that aged and experienced workers have more job satisfaction. Attitude also tends to be high when people enter the workforce. It has been found that when they enter the job, there is an unrealistic assumption towards what they are going, but they notice that reality fails for, short at their perceived expectations. So they will have a good attitude but not so well, in their median period. When they realize their disillusionment and get the accurate expectations because of which the job will be seen in a positive perspective resulting in higher attitude.

Race and Sex

Sex and race also affect the attitude toward their profession. In a study conducted by Quinn et al., in America, they found that job satisfaction among blacks and minority groups had been consistently lower than the whites. In a study by Charles N. Weanch, he found that there was no significant difference in job satisfaction when both male and female are equally affected by factors such as wages, prestige and supervisory positions.

Education Level

Education level plays an important role on attitude and job satisfaction. Researches conducted have revealed that when more educated persons are employed in low ranks; there is low job satisfaction and poor attitude toward their profession.

Personal Adjustments

Personal adjustments the teachers undergo in the society, family etc., also plays an important role in their attitude formation and change. The financial problems, the compulsions of employment, problems arising due to employment in the family, other familial problems are some of the personal problems they face thus leading to mental dissatisfaction or stress. Mental satisfaction and job satisfaction is considered as the major criteria for the performance of their job and their attitude towards it.

NEED FOR THE STUDY

Education is an important aspect in a person's overall development as an efficient being. Today education for children, though with up-to-date technological reach, has become uninteresting, due to the instructional materials and methods, lack of understanding, negative attitude towards the teacher, interest in some other activities, and some familial problems. It is also still unreachable for some students due to economic and social status of their families. The main reasons cited by the educationalists are teacher's attitude towards the children and the profession, and the teacher's frequent absence, inability to cope up with the children and work, thus resulting in stress, etc. A positive attitude, dedication, good health and a stress-free life are essential for a teacher to develop the knowledge and attitude of the students. If the attitude towards teaching is good and desirable, it will make the teaching, in turn learning effective.

In spite of their positive attitude and interest towards teaching, some of the teachers face stress in their job due to some reasons as lack of support from the institution and family. This may lead to a block in their commitment and the societal development. They must have a positive attitude, job satisfaction and less job stress, in order to give a wholesome involvement to the society. Assessing their attitude towards profession and the stress faced by them help us in identifying the causes and effect of the stress and low attitude, thereby helping them in their improvement and the wholesome improvement of the society. Identifying the level of stress and attitude among teachers would have been studied earlier, but the emphasis is given on the special educators in the present study. Assessment of their level of stress and attitude helps us in recognizing the main causal factors, and the remedies, thus to get their excellence in the job.

2

REVIEW OF RELATED RESEARCH

The review of scientific findings in every field of investigation shows significance of problems undertaken for the study. It is helpful in focusing the appropriation of methodology, procedure and analysis of data collected. It reveals the background of problem undertaken for the study to the investigator as well as the reader.

Durham (1976) in a survey of 658 teachers from Infant, Junior and Secondary schools identified the common stress situations, which teachers reported they had to cope with in their work. The three common stressors they identified were reorganization, role conflict and role ambiguity, and poor conditions. Sukhwal (1976) has found that married women teachers had a favourable attitude towards teaching profession.

Gupta (1979) found that women primary teachers have more favorable attitude than the men teachers towards teaching. Jalul and Pillay (1979) investigated the attitude of the college teachers towards teaching profession. They found that the age and experience of the college teachers have definite relationship with attitude towards teaching profession. It was also found that as teachers grew in age and experience their attitude towards teaching profession also grew more favourable. Further, it was noticed that the teachers who had developed sound attitude towards teaching profession were relatively more traditional and impersonal.

Kushwaha (1979) involved secondary school teachers of Rajasthan to find out their attitude towards children and schoolwork. It was found that the length of teaching experience was not an influencing factor of teaching attitude. The study revealed that the male teachers were better as advisers and disciplinarians whereas female teachers were better in motivating and counseling.

In a survey of 219 teachers in 16 medium-sized, mixed comprehensive schools in England, Kyriacou and Sutcliffe (1979) found that 14 significant associations between sources of stress and job satisfaction, frequency of absences and intention to leave teaching were all in the predicted directions. The source of stress, which was found by these authors to be mostly strongly associated with job satisfaction, frequency, absences and intention to leave teaching, was poor career structure and inadequate salary.

Gupta (1980) undertaken a study on the job satisfaction of the teachers at three levels—primary, secondary school and college. It was found that (i) attitude towards teaching and job satisfaction was positively related. (ii) Marital status, age and length of teaching experience were not significantly associated with job satisfaction in respect of primary and secondary teachers.

Singh and Dash (1980) have studied the attitude of male and female primary school teachers and they found that there was a significant relationship between ages; sex and teaching attitude with increasing age female teachers become more favorable towards teaching, whereas male teachers become less favorable.

Toni Amodio, Wayne State University (1981) studied on "An analysis of job-related stress and dissatisfaction in the teaching profession". They conducted a study on 181 teachers from urban and suburban school districts and found that 80 per cent of them suffered from occupational dissatisfaction, and the majority held rather negative attitude towards teaching as a career. The main three areas contributed for the decision were: Pupil characteristics, administrative competence and conflict, and professional isolation.

Rawat and Srivatsava (1984) studied the attitude of male and female teacher trainees towards teaching profession. For this purpose the researchers used T. A. inventory by Ahluwalia (1978) and found that there is significant difference between male and female trainees towards teaching profession.

Armes (1985) conducted a study on the sources of stress as reported by them and found that Head teacher support and salary were the main cause of dissatisfaction. He conducted the study on 291 teachers and found that 60 per cent of the respondents found their Head teacher to be the source of stress and only 44 per cent of them found their Head as a means of alleviating stress. He also found that among the teachers surveyed, 45.1 per cent reported dissatisfaction with their salary caused them some stress, 18.2 per cent suffered considerable stress through being dissatisfied with their salary and it is significant that 30 per cent of the sample experienced no stress arising from the considerations of their salaries.

Huling and Austin (1986) found that of all beginning teachers who enter the profession, 40-50 per cent leave their profession during their first seven years and in excess of two-thirds of those will do so in the first four years of teaching. First year teachers are 2.5 times more likely to leave the profession than their counterparts. An additional 15 per cent of beginning teachers will leave after their second year and still another 10 per cent will leave after the third year. They found that beginning teachers leave the teaching field is the inability to cope with teaching problems. Discipline, difficulties with parents and lack of sufficient or appropriate teaching materials are some of the problems experienced by beginning teachers.

Loadmen, *et al.* (1986) did a research on "Attitudes toward education and effectiveness of the classroom teacher, student teacher relationship". They obtained three independent ranking of classroom teacher's supervisory effectiveness from 47 classroom teachers, 91 student teachers and 12 university personnel. All were familiar with the supervisory behaviour of the 47 teachers. Each classroom teacher and student teacher also completed 2 standardized attitudes toward education scales which yielded 4 scores. Results indicate that teachers with highly programming or highly traditional attitudes towards education tended be perceived by student teacher as somewhat less effective supervisor. There were consistent differences between supervisory classroom teachers and student teachers with respect to attitude towards education.

Grissmer and Kirby (1987) conducted a study on teacher attrition in the field of special education and found that the attrition rate for both general and special educators followed a U-shaped curve, based on the age. He found that attrition rate is high among younger teachers, low for teachers during the mid-career period, and high again as teachers retire. The main reasons for their stress being salary, fewer debt obligations, and less invested in specific occupation.

Keith, *et al.* (1990) conducted a study on "Correlates of Psychological distress among secondary school teachers". They conducted the study on 574 Western Australian Secondary teachers, and found that the level of psychological stress was twice than that of the general population. Eight school-related factors were found to be statistically related to stress, such as inadequate access to facilities, lack of collegial support, perceived lack of achievement, excessive societal expectations, lack of influence or autonomy, student misbehavior, and lack of praise and recognition. They also found that the relationship between these factors and stress were stronger for female than the male teachers.

Mouli and Reddy (1990) have conducted a study on a sample of 100 teachers selected from 8 secondary schools located in Hyderabad and Secunderabad. They found that attitude of the teachers differ on the basis of sex, age and experience. But they concluded that the differences are not statistically significant.

Perkins (1991) observed that teacher satisfaction was not significantly affected by background variable such as teacher or principal gender, years of experience or school type assignment. He also found that teachers are most satisfied with their co-workers and least satisfied with the monetary aspects of teaching. Inadequate salary, low status of the profession and excessive paperwork are some common source of stress that affects the job satisfaction.

Cobb, *et al.* (1992) did a research on "An organizational behaviour analysis of teaching attitude about teaching". They used a general analysis of organizational behaviour theory to construct and test hypothesis concerning the determinants of teacher work attitudes. Results reveal that expectation of motivated students increased teacher's enthusiasm and they show more positive attitude towards teaching.

Richardson, *et al.* (1992) conducted a study of a methodological explanation of organization beliefs about teaching and their relationship to practice. The research was undertaken irrespective to a personal dilemma of the researchers about congruence between preferred beliefs and practice results show that teacher's belief about teaching is largely reviewed and shows play an important role in shaping their practice as teacher.

Brownell (1993) in his study on "Understanding special education teacher attrition: A conceptual model and implications for teacher educators" identified that the highest group of special education teachers at-risk for attrition are those with 5 years or less experience. The problems cited were discipline, problems with parents and lack of sufficient or appropriate materials.

Cooper and Kelley (1993) studied occupational stress amongst head teachers in the United Kingdom. They concluded that primary head teachers were experiencing higher levels of job dissatisfaction and stress than their secondary colleagues. The main sources of stress are work overload and handling relationships with staff.

Kirby and Grissmer (1993) studied data generated from 50,000 Indiana teachers over a 32-year period and found that half of the teacher population in the study left teaching by the end of fourth year. According to the survey, the primary reasons for the dissatisfaction of the teachers who left were; poor salary (45%), lack of student motivation (38%), inadequate administrative support (30%), student discipline problems (30%) and inadequate preparation time (23%).

Munn and Johnstone (1993) provided a snapshot of teacher's workload in schools. He conducted the study on 570 teachers selected from different sectors. They were asked to maintain a workload dairy and an Occupational Stress Indicator Questionnaire was to be answered. The response rate was 66 per cent for the dairy and 62 per cent for the Questionnaire. Over a typical week teachers recorded an average of 42.5 hrs of work. Over ninety three per cent (93%) reported at least one occasion when they felt stressed during the survey week. Significantly, the longer the hours worked, the more stress occasions were reported. Workload, new demands, administrative tasks and planning associated with change were identified as stressors.

Singer (1993) found that many teachers are leaving special education positions in favour of regular education positions, new careers and retirement. The largest among them are being the females below the age of 35 who have taught for less than 5 years. He found that young special educators leave at rates nearly twice than that of mature teachers.

Stancic (1993) studied teacher's attitudes toward teaching as a determinant of their readiness for Additional Professional Special Education. The sample of 98 teachers shows that teacher's decision to enroll in additional special education depends significantly on their attitude towards teaching. The teacher's satisfaction with own profession influences the success of integration as it is reasonable to expect that teachers satisfied with own profession will not view integration only as additional burden and obligation, but show readiness to accept it. Attitude towards teaching include the success of teacher-pupil relationship and the teachers' satisfaction of their profession.

Billingsley (1994) showed that younger special; educators are more likely to leave the profession than older special educators. Singer found that young special education teachers leave at rates nearly twice that of mature teachers. Boe, et al. showed that the age functions differently for leavers (those who exit public school teaching) than movers (those who changed positions). They decline systematically with increasing age.

In a study of 99 teachers who exited an urban school system, Billingsley et al. (1995) found that 37 per cent of the special educators compared to 53 per cent of general educators, left primarily for personal reasons, e.g., family or person move, pregnancy/child rearing, health, retirements.

Gonzalez (1995) conducted a study on "Factors that influence teacher attrition" and found that job stress leads special educators both novice and experienced, to switch to general education or to leave teaching altogether; the paperwork was often cited as the source. Teacher attitude and curriculum change, based on findings of this study and research by others, suggests that there is an important relationship between teacher's attitude and teacher's behaviour.

Tsuei (1995) conducted an analytic study of the relationship between and among principals and teacher's perception towards assumption of teaching and learning and integrated curriculum. The population for this study was all public school teachers in the state of Colorado. A stratified proportional sample of 1000 teachers was drawn from the population. Based on the analysis, the following conclusion was reached. There is no significant difference between elementary and middle teachers perception towards teaching.

Henke (1996) found that the majority of special education teachers (95%) work in public schools; a much smaller number (4%) work in private schools. Singh and Bellingsley (1996) found that the teachers who had higher levels of education, less experience, and belonged to a minority group were more likely to intend to leave because of better career alternatives outside of education. Westling and Whitten (1996) found that the special educators who were primary bread winners were more likely to stay than those who were not.

Boe (1997) did not find a relationship between gender and attrition for a national sample of general or special educators. However, in a study of urban special educators, Morvant, et al. (1995) found that male teachers are more likely to indicate intention to leave.

Ax, M and J. T. Stephens (1998) conducted a study on 237 randomly chosen Emotional and Behavioral disorder handling teachers in Wisconsin to explore the reasons these teachers enter the field and to investigate potential causes of their attrition. They were asked to complete a survey designed to identify the reasons for entering the field, and factors that may cause them to leave this field. Of those teachers surveyed, 42 per cent indicated lack of support as a primary reason. Twenty-five per cent highlighted the lack of administrative support, 15 per cent reported stress might justify leaving the filed and 13 per cent did not indicate a potential reason. Another 13 per cent indicated that student violence and 11 per cent indicated excessive paper work as the reason for leaving the field. The most frequent reason given by the respondents for entering the field was the desire for a challenge. Less frequently reported reasons included the lack of teaching positions in an original certification area and previous positive experiences with at-risk children.

Troman (1998) studied the consequences of stress on their personal lives. In a small-scale study of 24 teachers, he found that teachers reporting chronic stress were often involved in break-up of marital or personal relationships, caring for a dependent relative who was chronically ill, or had experienced the death of close relationship.

Boe, Barkanic and Leow (1999) have showed that there has been a "fairly high level of school attrition in the public school teaching force" with about 7 per cent of teachers moving to other positions and 6 per cent of them exiting teaching altogether for a total of 13 per cent. Special and general educators leave at about the same rates; however, special educators are significantly more likely than general educators to transfer to other teaching assignments.

Griffith, *et al.* (1999) questioned 780 primary and secondary school teachers, aiming to assess the associations between stress, coping responses and social support. High levels of stress were associated with low social support and the use of disengagement and suppression of competing activities as coping strategies.

Miller, *et al.* (1999) showed that teachers with less experience are more likely to leave and also indicate intent to leave more often than that of their more experienced counterparts. Soor (1999) investigated teacher's education and their attitude towards teaching and educational integration of children with special needs. In the study he scientifically founded selection of future teacher with special attention paid to the (a) personality traits (b) quality teacher education (c) continuing teacher training. The result shows that the teachers who form more relation with their pupils are more satisfied with their profession, which means that they have more positive attitude towards teaching.

Chen, *et al.* (2000) conducted a study on 1400 teachers, found that only small percentages of teachers were satisfied with the following aspects of their job: Only 26.5 per cent were satisfied with student's motivation to learn; 32.7 per cent were satisfied with student discipline and/or behaviour and 31 per cent were satisfied with parental support. Clearly, the handful of studies suggested that many teachers are dissatisfied with various aspects of their profession. They also conducted a study on 11,000 teachers and teacher candidates; found that only 14.6 per cent of the teachers surveyed were satisfied with the esteem in which society held the teaching profession.

In a National Survey of 1000 teachers, Farkas, Johnson and Foleno (2000) found that 66 per cent of the population they surveyed maintained that they did not feel respected and appreciated and 76 per cent argued that teachers are made scapegoats for all the problems in education.

Hastings (2002) identified challenging behaviors as a source of staff stress. In the study, 55 teachers and support staff in special schools for children with mentally retarded completed questionnaires assessing burnout, coping strategies for challenging behavior and their exposure to them. Results showed that (a) use of maladaptive coping strategies for challenging behaviours constitutes a risk for staff burnout, (b) this risk is in addition to that associated with exposure to challenging behavior and use of these strategies moderated the impact of exposure to challenging behaviours on emotional exhaustion burnout.

Nichols, *et al.* (2002) conducted a study on "burnout among special education teachers in self-contained cross-categorical classrooms" and found that special education teachers are leaving the field in much greater numbers than their peers in general education. They found lack of recognition for their work, by principals and other teachers as the main contribution to stress and burnout.

3

METHODOLOGY OF RESEARCH

The present study was conducted to assess the stress and attitude of women teachers working with normal and special children. This chapter deals with the methodology of the present study, herein the investigator has given the statement of the problem, the selection and the description of tools, the administration procedure and the description of the samples used.

STATEMENT OF THE PROBLEM

The selection of a problem is an initial step in conducting a research. Usually a number if problems are identified analyzed and evaluated before one of them is selected for research. When a suitable problem has been selected, its statement should be carefully given.

The statement of the problem taken for the present study of research is "a study of stress and attitude of women teachers working with normal and special children".

OBJECTIVES OF THE STUDY

Based on the conceptual ideas and established theories, the following objectives are framed for this present study.

1. To study the stress level of the women teachers handling normal and special children.

2. To study the attitude of the women teachers handling normal and special children towards their profession.
3. To study the level of stress, its control and the attitude of women teachers belonging to different age groups.
4. To study the level of stress, its control and attitude of women teachers with different years of experience.
5. To study the level of stress, its control and attitude of women teachers working under different management of schools.

HYPOTHESES OF THE STUDY

Having these objectives, the following null hypotheses were formulated. They are:

Hypothesis I

There would be significant difference in the level of stress among the women teachers handling normal and special children.

Sub-Hypotheses

1. There would be significant difference in the level of stress among the women teachers handling normal and visually challenged children.
2. There would be significant difference in the level of stress among the women teachers handling normal and hearing impaired children.
3. There would be significant difference in the level of stress among the women teachers handling normal and mentally challenged children.
4. There would be significant difference in the level of stress among the women teachers handling visually challenged and hearing impaired children.
5. There would be significant difference in the level of stress among the women teachers handling visually challenged and mentally challenged children.

6. There would be significant difference in the level of stress among the women teachers handling hearing impaired and mentally challenged children.

Hypothesis II

There would be significant difference in the level of attitude towards teaching profession among the women teachers handling normal and special children.

Sub-Hypotheses

1. There would be significant difference in the level of attitude towards teaching profession among the women teachers handling normal and visually challenged children.
2. There would be significant difference in the level of attitude towards teaching profession among the women teachers handling normal and hearing impaired children.
3. There would be significant difference in the level of attitude towards teaching profession among the women teachers handling normal and mentally challenged children.
4. There would be significant difference in the level of attitude towards teaching profession among the women teachers handling visually challenged and hearing impaired children.
5. There would be significant difference in the level of attitude towards teaching profession among the women teachers handling visually challenged and mentally challenged children.
6. There would be significant difference in the level of attitude towards teaching profession among the women teachers handling hearing impaired and mentally challenged children.

Hypothesis III

There would be significant difference in the level of stress among the women teachers of age below 40 years and above 40 years.

Hypothesis IV

There would be significant difference in the level of attitude towards teaching profession among the women teachers of age below and above 40 years.

Hypothesis V

There would be significant difference in the level of stress among women teachers with experience below 15 years and above 15 years.

Hypothesis VI

There would be significant difference in the level of attitude towards teaching profession among women teachers with experience below 15 years and above 15 years.

Hypothesis VII

There would be significant difference in the level of stress among teachers working under schools of different management.

Sub-Hypotheses

1. There would be significant difference in the level of stress among women teachers working in Government and Government Aided schools.
2. There would be significant difference in the level of stress among women teachers working in Government and Private Schools.
3. There would be significant difference in the level of stress among women teachers working in Government aided and Private Schools.

Hypothesis VIII

There would be significant difference in the level of attitude towards teaching profession among teachers working under schools of different management.

Sub-Hypotheses

1. There would be significant difference in the level of attitude towards teaching profession among women teachers working in Government and Government Aided schools.
2. There would be significant difference in the level of attitude towards teaching profession among women teachers working in Government aided and Private Schools.
3. There would be significant difference in the level of attitude towards teaching profession among women teachers working in Government aided and Private Schools.

SELECTION OF TOOLS

The tools used for the present study were Stress Questionnaire developed by Latha and Attitude Questionnaire constructed by Shalini Bhogle.

1. Stress Questionnaire

Rationale

This test was administered to adults as it was suitable to measure their level of stress. As it was standardized on Indian population it was more meaningful to use the scale for the present study.

Description

The scale consists of 52 statements arranged from mild stress (least affecting the everyday affairs), moderate to severe (which affects the adjustments and efficiency of an individual). This lists the life experiences based on the amount of "change" or "adjustments", one has to make to his/her life rather than the undesirability of events themselves. The items from 1 to 17 are classified as mild stressors, the items from 18-35 are moderate stressors and items ranging from 36-52 are classified as severe stressors.

It has a control index, where the subject has to record whether he or she have complete, partial or no control over the experienced stressful situation.

Scoring

The level of stress is scored by adding the "Yes" responses given by the respondent. The scoring under the level of stress category ranges from 0 to 52. To assess their control of stress, the scores are given as 1, 2 and 3 for the complete, partial or no control respectively.

Interpretation

Level of Stress

Scores	*Interpretation*
0 to 17	Mild Stress
18 to 35	Moderate Stress
36 to 52	Severe Stress

Control Index

Scores	*Interpretation*
0 to 52	Complete control of stress
53 to 104	Partial control of stress
105 to 156	No control of stress

2. Attitude Towards Teaching Profession Scale

Rationale

This test was used as the assessing tool as it can be administered to all age groups of teachers and was suitable to assess their attitude. It was an established, easy to apply tool, which consumes less expenditure, time and effort. As it was standardized on Indian population, it was more meaningful to use the scale for the present study.

Description

This tool is a questionnaire constructed by Shalini Bhogle. It consists of 18 items. These items list out the professional experiences based on the administration, curricula and the professional requirements. Each statement in the questionnaire is responded using a five-point scale, based on their agreeableness. They are strongly agree, agree, agree, undecided, disagree and strongly disagree.

Scoring

Scoring is based on summated ratings. Each response of the scale are scored as:

Response	*Strongly Agree*	*Agree*	*Undecided*	*Disagree*	*Strongly Disagree*
Score	1	2	3	4	5

The maximum possible score is 90.

Interpretation

Low level of scoring indicates a high level of job satisfaction and positive attitude towards teaching profession; High level of scoring indicates a low level of job satisfaction and thus negative attitude towards teaching profession.

Scores	*Interpretation*
18-42	High level of attitude
43-66	Moderate level of attitude
67-90	Low level of attitude

Pilot Study

A Pilot study was conducted to test the feasibility of the questionnaire and the adaptability of the samples. Getting permission from the school authorities, the researcher administered the test to 30 teachers and special educators as pilot study.

Administration Procedure

The main study was conducted among the selected samples in accordance with the variables such as disability handled by the teachers, their age, experience and the type of schools they are working. The researcher got prior permission from the management of the school authorities and discussed about the study conducted. The teachers were asked to be seated comfortably, the information such as name, age, qualification, educational and technical, experience were collected. Then the stress questionnaire was given to them with the following brief instructions. "Here is a list of events or situations or problems that is faced in day to day life. If you have experienced them and found them to be stressful, then, mark with a "yes" response and if not with a "no" response. It you find any of the situations not applicable to you, and then neglect them. Put a tick mark in the appropriate column under the control of index, based on your control over the situation". There was no time limit. Then the questionnaire was collected and they were asked to relax.

Later, the questionnaire for scoring attitude was administered to the individuals. They were asked to read the statements carefully and to mark in the appropriate column based on their agreeableness with the same. There was no time limit given. They took approximately twenty to thirty minutes to complete the questionnaires.

Reliability and Validity

Reliability

The Reliability is the degree of consistency that the instrument or procedure demonstrates whatever the tool is measuring. The split half reliability was used to find out the reliability of stress and attitude questionnaire. During the pilot study, this reliability was established. R = 0.76 and 0.79 respectively for both the tests.

Validity

Validity refers to the degree in which our test or other measuring device is truly measuring what we intended it to measure. Validity always refers to the degree to which that evidence supports the inferences that are made from the scores. It is the inferences regarding specific uses of a test that are validated, not the test itself. The items of questionnaire must appropriately sample a significant aspect of the purpose of the investigation.

The validity of the Stress Questionnaire used for the study is 0.86, taken for a sample of 80 subjects.

The validity of the attitude questionnaire was established by the author of the test. Hence the present researcher did not make any attempt to establish validity again.

SELECTION OF SAMPLE

Sampling is an important process used to select subjects for an experiment. Research results can be generalized only if the sample of participants studied represents that population accurately. When choosing a sample, psychologists must consider the possible impact that variables such as age, gender, race, ethnicity, cultural background, socioeconomic status, sexual orientation, disability, and so on can have on the behavior or mental process being studied.

For the present study, stratified random sampling method was used. The samples were selected, taking into consideration, the variables such as the age, experience, type of children they are handling and the type of school management they are working in.

In the present study for assessing the stress and the attitude of women teachers handling normal and special children, the sample size was 100.

Sample Distribution

Samples were distributed based on the variables such as age, experience, type of children handled by them, and the type of management they are working with. Thus they were distributed as the following:

Type of Children Handled

S.No.	*Nature of Respondents*	*No. of Sample*
1.	Women Teachers handling normal children	25
2.	Women Teachers handling visually challenged children	25
3.	Women Teachers handling hearing impaired children	25
4.	Women Teachers handling mentally challenged children	25

Age

S. No.	*Nature of Respondents*	*No. of Sample*
1.	Women Teachers of age below 40 years	45
2.	Women Teachers of age above 40 years	55

Experience

S. No.	*Nature of Respondents*	*No. of Sample*
1.	Women Teachers of experience below 15 years	63
2.	Women Teachers of experience above 15 years	37

Type of School Management

S. No.	*Nature of Respondents*	*No. of Sample*
1.	Women Teachers working in Government schools	30
2.	Women Teachers working in Government Aided schools	35
3.	Women Teachers working in Private Schools	35

DATA ANALYSIS

The data obtained from the sample of 100 women teachers working with normal and special children and analysed with the help of computer. The analysis involved mean, standard deviation, application of t-tests and analysis of variance.

RESULTS AND DISCUSSION

The following are the results of the present study on the stress and attitude of women teachers working with normal and special children. The results are followed by appropriate discussion.

1. Stress

Table 4.1

The level of stress among women teachers working different types of children (Normal and special children)

Groups	*N*	*Mean*	*Standard Deviation*	*F-Value*
Teachers handling normal children	25	20	4.13	3.314* Significant
Teachers handling visually challenged children	25	20.24	3.74	
Teachers handling hearing impaired children	25	17.76	2.52	
Teachers handling mentally challenged children	25	20.40	2.96	

* p= 0.05.

As per the selected stress questionnaire the score of 17 and above indicates a moderate level of stress. The above mean scores of all the teachers handling different special children and normal children fall within this category. Hence it can be interpreted as these women teachers are having moderate level of stress.

The above table shows the "f" value 3.31, which is significant at 0.05 level indicates that there is a significant difference in the level of stress among the teachers handling different special group of students and non-disabled students. This finding accepted the Hypothesis I, i.e., "There would be significant difference in the level of stress among the teachers handling normal and special children".

It is possible to say from this finding the level of stress is different among different categories of teachers. Further to find out which group of special teachers having more stress, the "t" test has been computed as follows.

This result supports the findings of Boe, Barkanic, and Leow (1999), which showed that the special educators are significantly more likely than the general educators to transfer to other teaching assignments. This result may be due to excessive paper work (Boe, Barkanic and Leow, 1999) challenging behavior of children they handle (Hastings, R.P. and Brown, T., University of Southampton, 2002) and inadequate support from administration, other professionals and colleagues (Ax, M and Stephens, J. T., 1998). Some of other reasons may be poor working conditions and facilities, poor salary, long hours of work, and heavy workload. Thus the level of stress of teachers handling special children is slightly more than that of the general teachers.

The Table 4.2 shows the "t" value of 0.22 which is not significant at any level. This finding indicates that there is no significant difference in the level of stress among the teachers handling normal and visually challenged children. So this finding rejected the Sub-Hypothesis I (a) "There would be significant difference in level of stress among teachers handling normal and visually challenged children".

Table 4.2

Comparison of the level of stress among women teachers handling normal and visually challenged children

Groups	*N*	*Mean*	*Standard Deviation*	*S.E*	*'t" value*	*Level of significance*
Teachers handling normal children	25	20	4.13	1.12	0.22	Not Significant
Teachers handling visually challenged children	25	20.24	3.74			

This result may be due to the more or less same classroom settings they work with. The challenging disabilities of children and sometimes the presence of children with multiple disabilities may be the cause of slightly high stress in teachers handling visually challenged children.

Some of the other reasons reported being excessive workload (Boe, Cook, Bobbit and Weber), more working hours and personal reasons or illness. Thus the stress level of teachers handling visually challenged children is more or less same with that of those handling normal children.

The table 4.3 shows the "t" value 2.31, which is significant at 0.05 level indicates that there is a significant difference in level of stress among teachers handling normal and hearing impaired children. This finding accepted the sub-hypothesis I (b), "There would be significant difference in the level of stress among the teachers handling normal and hearing impaired children".

This result may be due to the discipline problems of children, lack of sufficient and appropriate materials (Brownell, M. T and Smith, S. W, 1993), lack of student motivation and inadequate administrative support. Some of also reported reasons were poor salary, parent teacher relationship, the inadequate timing for meeting with parents, and long working hours. Thus the level of stress of teachers handling normal children is more than that of hearing impaired children.

The Table 4.4 shows the "t" value of 0.39, which is not significant at any level. This finding indicates that there is no significant difference in the level of stress among the teachers handling normal and mentally challenged children. This finding rejected the sub-hypothesis I (c), "There would be significant difference in the level of stress among the teachers handling normal and mentally challenged children".

Table 4.3

Comparison of the level of stress among women teachers handling normal and hearing impaired children

Groups	*N*	*Mean*	*Standard Deviation*	*S.E*	*"t"*	*Level of significance*
Teachers handling normal children	25	20	4.13	0.97	2.31	0.05 Not Significant
Teachers handling hearing impaired children	25	17.76	2.52			

Table 4.4

Comparison of the level of stress among women teachers handling normal and mentally challenged children

Groups	*N*	*Mean*	*Standard Deviation*	*S.E*	*"t"*	*Level of significance*
Teachers handling normal children	25	20	4.13	1.02	0.39	Not Significant
Teachers handling mentally challenged children	25	20.40	2.96			

This result does not supports the findings of Brownell, Smith, and Miller, 1995, who showed that stress was mostly reported by the teachers handling emotional and behavioral disorders children. This may be due to that the workload, paper work, working hours, the discipline problems in children, lack of children motivation, the frequent parent- teacher meetings, and time constraint for both the teachers being the same.

The presence of different degree of disability and the challenging behaviors may be the reason for the stress level being slightly high for teachers handling mentally challenged children.

The Table 4.5 shows the "t" value of 2.75, which is significant at 0.01 level indicates that there is a significant difference in the level of stress among the teachers handling visually challenged and hearing impaired children. So this finding accepted the Sub-Hypothesis I (d), " There would be significant difference among the women teachers handling visually challenged and hearing impaired children".

This result of teachers handling visually challenged children reporting more stressed may be mainly due to the challenging disability of the children they handle. Other reasons reported were excessive workload, the status of the profession in the society, more teaching assignments and additional administrative work, and time management.

The Table 4.6 shows the "t" value of 0.17, which is not significant at any level. This finding indicates that there is no significant difference in the level of stress among the teachers handling visually and mentally challenged children. This finding rejected the Sub-hypothesis I (e), "There would be significant difference in the level of stress among the teachers handling visually and mentally challenged children".

This result does not support the findings of George, George, Gersten and Grosenick (1995) and Brownell, Smith and Miller, 1995, who have showed that among the special educators, the teachers of students with emotional and behavioral disorders are more likely to transfer to other teaching assignments or leave the profession.

Table 4.5

Comparison of the level of stress among women teachers handling visually challenged and hearing impaired children

Groups	*N*	*Mean*	*Standard Deviation*	*S.E*	*"t" value*	*Level of significance*
Teachers handling visually challenged children	25	20.24	3.74	0.90	2.75	0.01 Not Significant
Teachers handling hearing impaired children	25	17.76	2.52			

Table 4.6

Comparison of the level of stress among women teachers handling visually challenged and mentally challenged children

Groups	*N*	*Mean*	*Standard Deviation*	*S.E*	*"t" value*	*Level of significance*
Teachers handling visually challenged children	25	20.24	3.74	0.95	0.17	Not Significant
Teachers handling mentally challenged children	25	20.40	2.96			

The result may be due to the reason that both the special educators work with children with varied disability. Some of other reasons reported were isolation from other faculty, lack of recognition of their work, lack of support from the parents and the administration and lack of the recognition for their profession.

The Table 4.7 shows the "t" value of 3.40, which is significant at 0.01 level indicates that there is a significant difference in the level of stress among the teachers handling hearing impaired and mentally challenged children. So this finding accepted the Sub-Hypothesis I (f), "There would be significant difference among the women teachers handling hearing impaired and mentally challenged children".

This result supports the findings of George, George, Gersten, and Grosenick (1995) and Brownell, Smith and Miller, 1995, who have showed that among the special educators, the teachers of students with emotional and behavioral disorders are more likely to transfer to other teaching assignments or leave the profession.

This result may be due to excessive paper work (Boe, Barkanic, and Leow, 1999), challenging behavior of children they handle, aggressive and physical abuse, lack of support from the administrators and other professionals, student violence (Ax. M. and J.T. Stephens, 1998). They may also be due to handling children with different degree of disability in the same class, more paper work, poor salary, health problems, and insufficient time to meet their responsibilities such as meeting their ward's parents or other professionals, which are being allowed after working hours.

This result also supports the findings of Brownell, Smith, and Miller (1995) who showed that stress was mostly reported by the teachers handling emotional and behavioral disorder's children among the special educators.

Thus, the stress level of teachers handling mentally challenged is found to be slightly higher than that of the teachers handling visually challenged children and hearing impaired children.

Table 4.7

Comparison of the level of stress among women teachers handling hearing impaired and mentally challenged children

Groups	*N*	*Mean*	*Standard Deviation*	*S.E*	*"t" value*	*Level of significance*
Teachers handling hearing impaired children	25	17.76	2.52	0.78	3.40	0.01 Not Significant
Teachers handling mentally challenged children	25	20.40	2.96			

Table 4.8

Attitude among women teachers handling different type of children (Normal and special children)

Groups	*N*	*Mean*	*Standard Deviation*	*F-value*
Teachers handling normal children	25	52.64	6.39	0.7809 Not Significant
Teachers handling visually challenged children	25	53.16	6.02	
Teachers handling hearing impaired children	25	51.32	5.98	
Teachers handling mentally challenged children	25	54	6.98	

As per the selected questionnaire for attitude, the score ranging 43 to 66 indicates a moderate level of attitude towards teaching profession. The above mean scores of all the teachers handling different special children and normal children fall within this category. Hence it can be interpreted as these women teachers are having moderate level of attitude towards their teaching profession.

2. Attitude

The Table 4.8 shows the "f" value 0.7809, which is not significant at any level, indicates that there is no significant difference in the level of attitude towards teaching profession among teachers handling different special group of students and non-disabled students. This finding rejected the Hypothesis II, i.e. "There would be significant difference in the level of attitude towards teaching profession among teachers handling normal and special children".

This result may be due to that all the teachers handling normal children and special children as the whole have the same perception of their professional requirements and constraints.

The result of the moderate level of attitude may be due to the monetary aspects of teaching (Perkins, 1991), poor working conditions, lack of motivation among students (Cobb, Steven and Foella, William, H., 1992), lack of support from the administration and colleagues and status given in the society for the profession.

The attitude towards teaching includes success of teacher-pupil relationship and the teacher's satisfaction in their profession. The lack of these main aspects may also the reason for moderate level of attitude among all the teachers. From the table, the level of attitude is found to be less, compared to others in the teachers handling mentally challenged children. This may be due to the handling of different levels of disability, challenging behaviors of children, frequent reporting and modification of curricula and more paperwork based on their ward's regular achievement.

Table 4.9

Comparison of level of stress among women teachers at different age groups

Groups	*N*	*Mean*	*Standard Deviation*	*t-value*
Below 40 yrs.	45	21.24	2.81	4.42* Significant
Above 40 yrs.	55	18.44	3.54	

* p = 0.01

3. Stress and Age

As per the selected stress questionnaire the score of 17 and above indicates a moderate level of stress. The above mean scores of the teachers of different age groups, falls within this category. Hence, it can be interpreted as these women teachers of different age group are having moderate level of stress.

The Table 4.9 shows the "t" value of 4.42, which is significant at 0.01 level, indicates that there is a significant difference in the level of stress among teachers of age below and above 40 years. This finding accepted the Hypothesis III, "There would be significant difference in the level of stress among the women teachers of age group below and above 40 years".

This result supports the study of Singer, 1993, who found that the largest among the teachers leaving the profession were below the age of 35 years and their attrition rate was twice than that of mature teachers. This finding also supports the study of Boe, Bobbit, Cook and Whitener (1997) which found that the young special education teachers leave at rates nearly twice than that of the mature teachers. This result also supports the findings of Grissmer and Kirby (1987) that the attrition rate is high among young special educators.

The reasons for the result may be due to the role conflict they face in the school, which includes not only teaching, paper work, assessment but also counseling the students, some administrative work, extra-curricular activities, etc. The young teachers enter the profession with more expectations about the profession, but when they are not met with, it leads to job stress. They being least satisfied with monetary aspects, are opened for many job opportunities, thus moving their jobs (Boe, Bobbit, Cook and Whitener, 1997). Other reasons reported were workload, handling relationship with other staff, challenging behaviours of children, and the recognition of the profession.

4. Attitude and Age

As per the selected questionnaire for attitude the score ranging from 43 to 66 indicates a moderate level of attitude towards teaching profession. The above mean scores of the teachers of different age groups, falls within this category. Hence, it can be interpreted as these women teachers of different age group are having moderate level of attitude towards teaching profession.

Table 4.10

Comparison of attitude of women teachers of different age groups

Groups	*N*	*Mean*	*Standard Deviation*	*t-value*
Below 40 yrs.	45	55.60	5.46	4.43* Significant
Above 40 yrs.	55	50.47	6.10	

* **p= 0.01**

The Table 4.10 shows the "t" value of 4.43, which is significant at 0.01 level, indicates that there is a significant difference in the level of attitude towards teaching profession among teachers of age below and above 40 years. This finding accepted the Hypothesis IV, "There would be significant difference in the level of attitude towards teaching profession among the women teachers of age group below and above 40 years".

This result supports the findings of Jalul and Pillay (1979) that as teachers grew in age their attitude towards their teaching profession also grew more favorable.

This result does not support the findings of Mouli and Reddy (1990) who found from a study of 100 secondary school teachers that the attitude of teachers toward their profession does not significantly differ on the basis of age and experience.

The result may be due to more expectations over the job, more workload, more working hours off and on the work, inadequate salary, poor career structure, less motivation in them and among the students, other familial responsibilities, and many job opportunities outside teaching with adequate salary.

6. Stress and Experience

As per the selected stress questionnaire the score of 17 and above indicates a moderate level of stress. The above mean scores of the teachers of different years of experience fall within this category. Hence, it can be interpreted as these women teachers of different experience level are having moderate level of stress.

Table 4.11

Comparison of the level of Stress among women teachers with different levels of experience

Groups	*N*	*Mean*	*Standard Deviation*	*t-value*
Below 15 yrs.	63	20.94	3.21	5.26 Significant
Above 15 yrs.	37	17.59	2.98	

* p = 0.01

The Table 4.11 shows the "t" value of 5.26, which is significant at 0.01 level, indicates that there is a significant difference in the level of stress among teachers of experience below and above 15 years. This finding accepted the Hypothesis V, "There would be significant difference in the level of stress among the women teachers of experience below and above 15 years".

This result supports the findings of Huling and Austin (1986) who found that out of all the beginning teachers entering the profession, 40-50 per cent leave their profession during the first seven years. This study also supports the findings of Singh and Billingsley, 1996 that the teachers who had higher level of education, less experience, were more likely to intend to leave the profession. It also supports the study of Brownell et al. (1993) that identified the highest group of special education teachers at risk being those with 5 years or less experience.

This result may be due to the lack of discipline and behavior problems in children, problems with parents, lack of sufficient or appropriate materials, workload, and inadequate support from the counterpart and other staff. They develop negative attitude toward their profession as they enter it with high expectations that are not met with in reality. Another reason cited was ending up with health problems due to the heavy working hours and work load.

7. Attitude and Experience

As per the selected questionnaire for attitude the score ranging from 43 to 66 indicates a moderate level of attitude towards teaching profession. The above mean scores of the teachers of different level of experience fall within this category. Hence, it can be interpreted as these women teachers of different experience levels are having moderate level of attitude towards teaching profession.

Table 4.12

Comparison of the level of attitude among women teachers with different levels of experience

Groups	*N*	*Mean*	*Standard Deviation*	*t-value*
Below 15 yrs.	63	55.25	5.61	6.02* Significant
Above 15 yrs.	37	48.57	5.22	

* **p = 0.01**

The Table 4.12 shows the "t" value of 6.02, which is significant at 0.01 level, indicates that there is a significant difference in the level of attitude towards teaching profession among teachers of experience below and above 15 years. This finding accepted the Hypothesis VI, "There would be significant difference in the level of attitude towards teaching profession among the women teachers of experience below and above 15 years".

This result supports the findings of Jalul and Pillay (1979) that the age and experience of the teachers have definite relationship with attitude towards teaching profession. It also supports the findings of Singer, 1993, who found that the largest among the teachers leaving the profession, are those who have taught for less than 5 years.

This study does not support the findings of Kushwaha (1979) who found that the length of teaching experience was not an influencing factor of teaching attitude.

The results may be mainly due to the low salary paid and low status of the profession in the society. The beginning teachers are opened to many job opportunities with attractive salaries and facilities. They have reported to have made the scapegoats for all the problems in the education (Farkas, Johnson and Foleno, 2000). Thus they develop a negative or moderate attitude toward their profession. Some of other reasons reported were low motivation of students, low administrative support, workload, physical and mental health problems due to immatureness and inability to handle with all responsibilities.

8. Stress and Different Managements of Schools

As per the selected stress questionnaire the score of 17 and above indicates a moderate level of stress. The above mean scores of the teachers working under different management fall within this category. Hence, it can be interpreted as these women teachers working under different management are having moderate level of stress.

Table 4.13

The level of stress among women teachers working in schools under different management

Groups	*N*	*Mean*	*Standard Deviation*	*F-value*
Teachers working in Government Schools	30	19.53	3.94	4.428* Significant
Teachers working in Government Aided Schools	35 35	18.57 20.97	3.08 3.19	
Teachers working in Private Schools				

* p = 0.05.

The Table 4.13 shows the "f" value 4.428 which is significant at 0.05 level indicates that there is a significant difference in the level of stress among the teachers working in schools of different managements. This finding accepted the Hypothesis VII, i.e., "There would be significant difference in the level of stress among the teachers working in schools of different managements".

It is possible to say from this finding the level of stress is different among teachers working different managements. Further to find out teachers under which management are having more stress, the "t" test has been computed as follows.

This study supports the findings of Bobbit, Faupel, Burns (1991) that the attrition rate for the teaching profession was 5.6 per cent in the Public schools and 12.7 per cent in the Private schools. This result supports the findings of Henke, Choy, Geiss and Boughman (1996) who found that the majority of teachers, about 95 per cent of their sample are working in public schools and 4 per cent of them were working in private schools.

This result may be due to more administrative jobs given, long working hours even after the normal school time, frequent and regular reporting to the school head and the parents, and inadequate support from the colleagues. Low status of esteem about the profession in the society due to the salary paid is also reported to cause stress in teachers. Pressure from parents and the administration about the children's academic and social improvement, extra-curricular activities, etc are also some of the cited reasons for stress.

The Table 4.14 shows the "t" value of 1.08, which is not significant at any level, indicates that there is no significant difference in the level of stress among teachers working in Government and Government aided schools. This finding rejected the Hypothesis VII (a), "There would be significant difference in the level of stress among the women teachers working in Government and Government aided schools".

Table 4.14

Comparison of the level of stress among women teachers working in Government and Government aided schools

Groups	*N*	*Mean*	*Standard Deviation*	*S.E*	*'t" value*	*Level of significance*
Teachers working in Government Schools	30	19.53	3.94	0.89	1.08	Not Significant
Teachers working in Government Aided schools	35	18.57	3.08			

This result may be due to the more or less same conditions of the school environment they work in. The government aided school teachers are mostly governed under the same terms and conditions, and paid with more or less same amount of perks as that of teachers working in Government. The level of stress reported by the government teachers are said to be slightly high than that of Government aided teachers. Most of the reasons reported were the number of students present under each faculty, the behavior and the economic background of the children, and more amount of paperwork and administration work demanded nowadays.

The Table 4.15 shows the "t" value of 1.60, which is not significant at any level, indicates that there is no significant difference in the level of stress among teachers working in Government and Private schools. This finding rejected the Hypothesis VII (b), "There would be significant difference in the level of stress among the women teachers working in Government and Private".

This result may be due to more or less same conditions faced by both the group of teachers such as workload, the student-teacher ratio and more amount of administrative work, the student's discipline problems like violence (Ax, M. Stephens, J.T., 1998) and inadequate motivation of students (Kirby and Glessmer, 1993), inadequate support from parents and inadequate time to spend for their personal aspects.

The Table 4.16 shows the "t" value of 3.20, which is significant at 0.01 level, indicates that there is a significant difference in the level of stress among teachers working in Government aided and Private schools. This finding accepted the Hypothesis VII (c), "There would be significant difference in the level of stress among the women teachers working in Government aided and private schools".

Table 4.15

Comparison of the level of stress among women teachers working in Government and Private Schools

Groups	*N*	*Mean*	*Standard Deviation*	*S.E*	*'t"*	*Level of significance*
Teachers working in Government Schools	30	19.53	3.94	0.90	1.60	Not Significant
Teachers working in private schools.	35	20.97	3.19			

Table 4.16

Comparison of the level of stress among women teachers working in Government Aided and Private schools

Groups	*N*	*Mean*	*Standard Deviation*	*S.E*	*'t"*	*Level of significance*
Teachers working in Government Aided Schools	35	18.57	3.08	0.75	3.20	0.01 Not Significant
Teachers working in Private Schools	35	20.97	3.19			

This result of high level of stress reported by private school teachers may be mainly due to the administration of the school being autocratic, decisions not being supportive or consultative. Some of other reasons reported were lack of recognition and appreciation of their work, low perks, relatively more work, and the low status of the profession. Some also reported the inadequate learning materials, and also personal financial problems and familial problems as the causal factor of their stress.

9. Attitude and Different Management Schools

Table 4.17

The level of attitude among women teachers working in schools under different management

Groups	*N*	*Mean*	*Standard Deviation*	*F-value*
Teachers working in Government Schools	30	50.37	4.91	7.023* Significant
Teachers working in Government Aided Schools	35	51.91	5.63	
Teachers working in Private Schools	35	55.71	7.06	

* $p = 0.01$

As per the selected questionnaire for attitude the score ranging from 43 to 66 indicates a moderate level of attitude towards teaching profession. The above mean scores of the teachers working in schools of different management fall within this category. Hence, it can be interpreted as these women teachers working in schools of different management are having moderate level of attitude towards teaching profession.

The Table 4.17 shows the "f" value 7.023 which is significant at 0.01 level indicates that there is a significant difference in the level of attitude towards teaching profession among the teachers working in schools of different managements. This finding accepted the Hypothesis VIII, i.e., "There would be significant difference in the level of attitude towards teaching profession among the teachers working in schools of different managements".

It is possible to say from this finding the level of attitude is different among teachers working different managements. Further to find out teachers under which management are having more positive attitude, the "t" test has been computed as follows.

The Table 4.18 shows the "t" value of 1.18, which is not significant at any level. This findings indicates that there is no significant difference in the level of attitude among the teachers working in Government and Government aided schools. So this finding rejected the Hypothesis VIII (a), "There would be significant difference in the level of attitude towards their teaching profession among the women teachers working in Government and Government aided schools".

This study does not support the findings of Henke, Choy, Geiss and Broughman (1996) that the majority of special education teacher work in public schools and only a much smaller amount work in private schools.

This result may be due to the terms and conditions of the administration, remunerations paid, incentives and promotions provided are more or less same as that of Government teachers. The level of attitude towards their profession being moderate may be due to more administrative jobs as equal to teaching jobs, the student's behaviors and their economical background, etc.

Table 4.18

Comparison of the level of attitude among women teachers working in Government and Government aided schools

Groups	*N*	*Mean*	*Standard Deviation*	*S.E*	*"t"*	*Level of significance*
Teachers working in Government Schools	30	50.37	4.91	1.31	1.18	Not Significant
Teachers working in Government Aided schools.	35	51.91	5.63			

Table 4.19

Comparison of the level of attitude among women teachers working in Government and Private schools

Groups	*N*	*Mean*	*Standard Deviation*	*S.E*	*"t"*	*Level of significance*
Teachers working in Government Schools	30	50.37	4.91	1.49	3.58	0.01 Not Significant
Teachers working inPrivate schools	35	55.71	7.06			

The Table 4.19 shows the "t" value of 3.58, which is significant at 0.01 level indicates that there is a significant difference in the level of attitude among the teachers working in Government and Private schools. So this finding accepted the Hypothesis VIII (b), "There would be significant difference in the level of attitude towards their teaching profession among the women teachers working in Government and Private schools".

The Table 4.20 shows the "t" value of 2.49, which is significant at 0.01 level indicates that there is a significant difference in the level of attitude among the teachers working in Government aided and Private schools. So this finding accepted the Hypothesis VIII (c), "There would be significant difference in the level of attitude towards their teaching profession among the women teachers working in Government aided and Private schools".

This study supports the findings of Bobbit, *et al.* (1991) that the attrition rate for the teaching profession was 5.6 per cent in the Public schools and 12.7 per cent in the Private schools. Thus there being a important relationship between stress and attitude.

The result of negative or low attitude towards teaching reported in private school teachers may be due to the autocratic type of management, the decisions made without consultations from the teachers, frequent and regular reporting to the management, low perks, and sufficient job opportunities outside teaching field with adequate salary, etc.

The pressure from the parents in the nowadays technological world, the pressure from schools to acquire best academic achievement, results in teachers being stressed and thus their attitude towards their profession deteriorates leading to quitting the job or working with the same attitude.

Table 4.20

Comparison of the level of attitude among women teachers working in Government Aided and Private schools

Groups	*N*	*Mean*	*Standard Deviation*	*S.E*	*"t"*	*Level of significance*
Teachers working in Government Aided Schools	35	51.91	5.63	1.53	2.49	0.01 Not Significant
Teachers working in Private Schools	35	55.71	7.06			

5

SUMMARY AND CONCLUSIONS

It is now generally accepted that stress is a multidimensional and a complex phenomenon, which is influenced by personal, situational and structural factors. Specifically stress in teaching is a multi level phenomenon, which results in unpleasant negative emotions such as anger, frustration, depression, etc., further affecting the profession.

Attitude is a generalized feeling or evaluative reaction to a particular object or profession, determined through psychological or socio-cultural factors. Attitude towards the teaching profession determines the level of job satisfaction they possess and thus their proficiency in teaching.

The present research study conducted to assess the level of stress and attitude of women teachers working with normal and special children. To measure the level of stress and attitude, the investigator administered the questionnaire method for data collection. The researcher selected the suitable tools for her research. The tools were Stress Questionnaire by Latha (1997) and Attitude Questionnaire by Shalini Bhogle. These tools were used to assess the level of stress and attitude respectively. The data was collected from a sample of 100 women teachers from the normal and special schools in the city, varied in the disability handling, age, experience, and management of the school they worked in.

The appropriate statistical analyses used to test the formulated hypothesis were analysis of variance and t-test. From the data analyzed, the following findings were revealed and the conclusions were drawn out of this research.

FINDINGS OF THE STUDY

The major findings of this study are the following.

Stress and attitude of women teachers handling normal and special children, reveals that the stress of women teachers as a whole is of moderate level, but those of teachers who are handling mentally challenged children is found to be relatively high compared with that of teachers handling other children.

The attitude level of the teachers is found to be moderate and there again the mentally challenged teachers are reporting relatively low level of attitude towards their profession.

The teachers handling hearing impaired children are found to have relatively low level of stress and high attitude towards their profession.

Teachers of age group of below 40 years and of experience below 15 years are found to have more stress and low attitude towards their profession.

Moderately high level of stress and low level of attitude is found in the teachers working under private management schools.

CONCLUSIONS OF THE STUDY

The research provides valuable insight into the field of teaching and the women teachers working with normal and special children. More importantly highlighted reasons for stress by the teachers were workload, challenging behaviors of the children, inadequate administrative support, differently degree of disabled children in a class, student—teacher ratio and time constraint. Moreover, the stress being an important psychological variable, experienced in day to day life, by everyone, causing psychological and physical health problems, is an additional cause for the teachers due to their professional requirements.

The level of attitude being inversely related to stress, those teachers with high level of stress were found to have low or negative attitude toward their profession. The main reason for negative attitude towards their profession was the lack of recognition of their work, low professional status and low remunerations paid.

In case of the young and less experienced teachers and special educators, the high level of stress and low level of attitude may be mainly due to the high expectations they possess on the profession, low remunerations, and more job opportunities outside the field of education. The stress level is optimally high and the attitude level is moderately low in teachers working in private schools, due to the lack of esteem in the society, low remunerations and administration constraints.

Thus the stress and low level of attitude reported results in great attrition rate among these teachers. If there is retention, it results in health problems, leading to frequent absenteeism, and thus their efficacy in teaching decreases.

The teachers play an important role in spreading education and building up a healthy society. Teachers play an important role in not only the academic development, but also in the behavioural, social and emotional development of children. The teachers in this process are prone to long hours of work, more commitments with the teaching and other professional requirements, and excessive workload. The teachers also have to struggle hard to prove their excellence, as their achievements are not being recognized or appreciated.

Thus, teaching as a profession requires a healthy amount of dedication, energy and time spent to it, to reach up to its level of excellence. Though they thrive towards path of excellence, they meet with many constraints as inadequate administrative support, time mismanagement, physical and mental health problems, thus ending up with low level of attitude and more amount of stress. Thus, due to more stress and low level of attitude, the teacher's effective teaching gets affected; it leads to frequent absenteeism, and also health problems such as headache, backache, spinal problems and cardiovascular problems. Hence all the above results indicate that there is moderate level of stress and attitude towards their profession in women teachers handling normal and special children.

IMPLICATIONS OF THE STUDY

This study could be used to create awareness among the schools and thus would immensely prevent losing effective teachers.

The finding would be helpful to the schools to follow an early intervention programs on the stress management for their teachers.

This study could also be used as a guideline to find the causal factors of the stress and negative attitude in related professions.

This finding could be helpful to the educational departments thus framing the teacher training curricula.

LIMITATIONS OF THE STUDY

The following were the limitations of the study.

The study involved women teachers only from urban areas.

The samples involved only the teachers handling visually challenged, hearing impaired and mentally challenged.

The variables used were only age, experience, and the type of management of the schools.

SUGGESTIONS FOR FURTHER RESEARCH

The following were some of the useful suggestions for future researchers in the area of the present study:

It could be done on women teachers working in schools present in other kinds of schools.

The same study could be done on a larger sample covering a total district or a state.

Stress management programs can be given and follow-up could be done for helping the teachers.

The study could be carried out for the teachers handling physically disabled children.

BIBLIOGRAPHY

Allinder, R. (1994), "Use of Time by Teachers in Various Types of Special Education Programmes". *Special Services in the Schools,* 9 (1), 125-136.

Anne C. Schafer (December 2001), "Demographic Overview: Teacher Workload Issues and Stress Survey", Spring 2001—BCTF Research Report (WLC-05).

Antonio, A. S., University of Manchester, Polychroni. F, University of Wales Athens Campus and Walters, B. University of Manchester, "Sources of Stress and Professional Burnout of Teachers of Special Educational Needs in Greece", Paper Presented at ISEC 2000.

Ax, Mary, Conderman Greg, Stephens, J. Todd., "Principal support essential for retaining special educators", National Association of Secondary School Principals, *NASSP Bulletin,* January 2001.

Ax, M., and J.T. Stephens, 1998, "Attrition and Competencies of Teachers who Teach Students with Emotional or Behavioural Disorders. Paper Presented at the Spring Tonic Teachers Conference, Manitowoc, University of Wisconsin, April.

Billingsley. B. S., (1993), "Teacher Retention and Attrition in Special and General Education: A Critical Review of the Literature", *Journal of Special Education*; 27(2); 137-174.

Biranchi Narayan Dash and Kunjalatha Dash, (2002), *Essentials of Educational Psychology,* Neelkamal Publications.

Boe. E.E., Bobbit. S.A., Cook. L. and Weber. A. L., (May, 1995), "Retention, Transfer and Attrition of Special and General Education Teachers in National Perspective". Paper Presented at the National Dissemination Forum on Issues Relating to Special Education Teacher Satisfaction, Retention and Attrition, Washington, D.C.

Carsten, J. M. and Spector, P.E. (1987), Unemployment, Job Satisfaction and Employee Turnover: A Test to the Muchinsky Model, *Journal of Applied Psychology,* 72, 374-381.

Charles E. Skinner, *Educational Psychology,* 4th Edition, Prentice-Hall India.

Charlie Naylor, "Analysis of Qualitative Data from the BCTF Work Life of Teachers Survey Series, Workload and Stress". BCTF Research, WLC – 03, October, 2001.

Cooper, C. and Payne, R. (1978), *Stress at Work,* London: John Wiley.

Cooper, C. L. and Kelley, M. (1993), "Occupational Stress in Head Teachers: A National UK Study". *British Journal of Educational Psychology,* 63, 130-143.

Dewe, P.J., Guest, D. and Williams, R. (1979), "Methods of Coping with Work-related Stress" in Mackay and Cox (eds) *Response to Stress: Occupational Aspects.* IPC Science and Technology Press.

Dunham, J. (1984 b), *Stress in Teaching,* Croom Helm.

Glenn, J., (2000), *Before it's too Late: A Report to the Nation from the National Commission on Mathematics and Science Teaching for the 21st Century,* Washington, DC; U.S. Department of Education.

Gold, Y. and Roth, R.A. (1993), *Teachers Managing Stress and Preventing Burnout: The Professional Health Solution,* Washington, D. C, Falmer Press.

Gonzalez, P., (August, 1995), Factors that Influence Teacher Attrition (NSTEP Information Brief 231-95). Alexandria, VA: National Association of State Directors of Special Education.

Griffith, J., Steptoe, A., and Cropley, M. (1999), "An Investigation of Coping Strategies Associated with Job Stress in Teachers, *British Journal of Educational Psychology*, 69, 4, 517-531.

Henke, R. R., Chen, X., Geis, S. and Knepper, P. (2000), "Progress through the Teacher Pipeline: 1992-93 College Graduates and Elementary and Secondary School Teaching as of 1997". (NCES Report No. 2000 – 152), Washington, DC: U.S. Department of Education., National Center for Educational Statistics.

Janice Croasmun, Donald Hampton and Suzannah Hermann, "Teacher Attrition: Is Time Running Out", Educational Leadership Program, School of Education, University of North Carolina.

Johnstone, M, (1993a), *Stress in Teaching: A Return to the Questions*, Edinburgh: SCRE

Johnstone, M, (1989), *Stress in Teaching: An Overview of Research*, Edinburgh: SCRE.

Kedjidjian, C. B. (1995), "Say No to Booze and Drugs in your Workplace". *Safety and Health*, 152 (6), 38-42.

Keith. F. Punch and Elizabeth Tuettemann, "Correlates of Psychological Distress Among Secondary School Teachers", *British Educational Research Journal*, Vol. 16, No. 4 (1990), pp. 369-382.

Kirby, S.N and Grissmer, D.W. (1993), *Teacher Attrition: Theory, Evidence, and Suggested Policy Options*, Santa Monica, CA: Rand Corporation. (ERIC DRS No: ED364533)

Kyriacou, C, (2001), "Teacher Stress: Directions for Future Research". *Educational Review*, 53, 1, 27-35.

Latha, S. (1997). Development of Stressful Life Events Questionnaire. *Journal of Psychometry*. Vol. IV, No. 2.

Lazarus, R. S. (1991), "Psychological Stress in the Worklplace" in E. Perrewe (Ed.), Handbook of Job Stress [Special Issue]. *Journal of Social Behaviour and Personality*, 6 (7), 1-13.

Leemamol Mathew, *An Exploratory Study on Occupational Stress and Coping Strategies of Special Educators in South India*, University of Calicut, January 2005.

Leah P. McCoy, (2003, March 28), "It's a Hard Job: A Study of Novice Teacher's Perspectives on Why Teachers Leave the Profession", *Current Issues in Education* [On-line], 6 (7).

Madden, H.D., Flanigan, J.L., Richardson, M.D. (1991, November), "Teacher Absences: Are there Implications for Educational Restructuring?", Paper Presented at the Annual Meeting of the Mid – South Educational Research Association, Lexington, KY (ERIC Document Reproduction Service No. ED 343 199)

Mary Brownell, (June 1997), "Coping with Stress in Special Education Classroom: Can Individual Teachers More Effectively Manage Stress?", University of Florida, (ERIC EC Digest No. 3545).

Mouli, R.C. and Reddy, S.V.B., "Attitude of Teachers Towards Teaching Profession", *Experiments in Education*, Vol. 18, 1990.

Selye, Hans. (1956), *The Stress of Life*, New York, McGraw - Hill.

Sherri Jennings Otto, Mitylene Arnold, (June 2005), *A Study of Experienced Special Education Teacher's Perception of Administrative Support*, Texas A and M University – Kingsville.

Skrtic, T.M., (1991), *Behind Special Education: A Critical Analysis of Professional Culture and School Organization*, Denver Love Publications.

Sundararajan, S. and Sabesan, "Job of High School Teachers in Relation of Their Attitude Towards Teaching and Their Job Involvement", *Experiments in Education*, Vol. 16, 1988.

Sundararajan and Vivekanandam. R, "Job Satisfaction of Teachers Working in Some Scheduled Higher Secondary School", *Experiments in Education*, Vol. 27, No. 12, 1990.

Todd W. Busch, "Teaching Students with Learning Disabilities: Perception of a First-year Teacher", *The Journal of Special Education*, Vol. 35, No. 2/2001/ pp. 92-99.

Toni Amodio, (1981), *An Analysis of Job-related Stress and Dissatisfaction in the Teaching Profession*, Wayne University.

Travers. C.J. and Cooper. C. L., (1996), *Teachers Under Pressure: Stress in the Teaching Profession*, London, Routledge.

Troman, G. (1998), "Living at a Hundred Miles an Hour: Primary Teacher's Perceptions of Work and Stress". Paper Presented at the British Educational Research Association Annual Conference, Queen's University, Belfast, August 27-30.

Valerie Wilson, (July 2002), *Feeling the Strain: An Overview of Literature on Teacher's Stress*, University of Glasgow, SCRE.

Vimala, T.D, Prasad Babu, B. and Bhaskara Rao, D. (2007), *Stress, Coping and Management*, Sonali Publications, New Delhi.

Zoe Ann Brown and Denise L. Uehara (1999), *Coping with Teacher Stress: A Research Synthesis for Pacific Educators*, Pacific Resources for Education and Learning, Honolulu.

REFERENCES

Bhaskara Rao, Digumarti (1994). *Scientific Aptitude*. New Delhi: Ashish Publishing House. ISBN 81-7024-658-X.

Bhaskara Rao, Digumarti (1995). *Animal Kingdom*. New Delhi: Discovery Publishing House. ISBN 81-7141-274-2.

Bhaskara Rao, Digumarti (1995). *Batracology*. New Delhi: Discovery Publishing House. ISBN 81-7141-279-3.

Bhaskara Rao, Digumarti (1997). *Scientific Attitude*. New Delhi: Discovery Publishing House. ISBN 81-7141-381-1.

Bhaskara Rao, Digumarti (1996). *Scientific Attitude vis-à-vis Scientific Aptitude*. New Delhi: Discovery Publishing House. ISBN 81-7141-308-0.

Bhaskara Rao, Digumarti (2004). *Scientific Attitude, Scientific Aptitude and Achievement*, New Delhi: Discovery Publishing House. ISBN 81-7141-781-7.

Bhaskara Rao, Digumarti (2004). *Educational Administration* New Delhi: Discovery Publishing House. ISBN 81-7141-842-2.

Bhaskara Rao, Digumarti (2004). *Issues in School Education*. New Delhi: Discovery Publishing House. ISBN 81-8356-025-3

Bhaskara Rao, Digumarti, Editor (1996). *Encyclopaedia of Education For All*, 5 Volumes. New Delhi: APH Publishing Corporation. ISBN 81-7024-759-4.

Vol. I *Education For All: The World Conference*. ISBN 81-7024-760-8.

Vol. II *Education For All: The EPA-9 Summit*. ISBN 81-7024-761-6.

Vol. III *Education For All: Quality Education For All*. ISBN 81-7024-762-6.

Vol. IV *Education For All: Planning and Monitoring*. ISBN 81-7024-763-4.

Vol. V *Education For All: The Indian Scenario*. ISBN 81-7024-764-0.

Bhaskara Rao, Digumarti, Editor (1996). *National Policy on Education*, 2 volumes. New Delhi: Anmol Publications Pvt. Ltd. ISBN 81-7488-323-1.

Bhaskara Rao, Digumarti, Editor (1996). *Global Perceptions on Peace Education*, 3 Volumes. New Delhi: Discovery Publishing House. ISBN 81-7141-319-6.

Bhaskara Rao, Digumarti, Editor (1997). *Education for the 21st Century*. New Delhi: Discovery Publishing House. ISBN 81-7141-389-7.

Bhaskara Rao, Digumarti, Editor (1997). *Reflections on Scientific Attitude*. New Delhi: Discovery Publishing House. ISBN 81-7141-319-6.

Bhaskara Rao, Digumarti, Editor (1997). *Success Story of a Primary Education Project*, New Delhi: APH Publishing Corporation. ISBN 81-7024-850-7.

Bhaskara Rao, Digumarti, Editor (1997). *World Food Summit*. New Delhi: Discovery Publishing House. ISBN 81-7141-386-2.

Bhaskara Rao, Digumarti, Editor (1997). Care the Child, 2 Volumes. New Delhi: Discovery Publishing House. ISBN 81-7141-394-3.

Bhaskara Rao, Digumarti, Editor (1998). *Earth Summit*, 2 Volumes. New Delhi: Discovery Publishing House. ISBN 81-7141-435-4.

Bhaskara Rao, Digumarti, Editor (1998). *Adolescence Education*. New Delhi: Discovery Publishing House. ISBN 81-7141-432-X.

Bhaskara Rao, Digumarti, Editor (1998). *Community and School Nutrition Education*. New Delhi: Discovery Publishing House. ISBN 81-7141-435-4.

Bhaskara Rao, Digumarti, Editor (1998). *District Primary Education Programme*. New Delhi: Discovery Publishing House. ISBN 81-7141-396-X.

Bhaskara Rao, Digumarti, Editor (1998). *National Policy on Education: Towards an Enlightened and Humane Society*. New Delhi: Discovery Publishing House. ISBN 81-7141-426-5.

Bhaskara Rao, Digumarti, Editor (1998). *Reforming School Education*. New Delhi: Discovery Publishing House. ISBN 81-7141-403-6.

Bhaskara Rao, Digumarti, Editor (1998). *Teacher Education in India*. New Delhi: Discovery Publishing House. ISBN 81-7141-406-0.

Bhaskara Rao, Digumarti, Editor (1998). *World Summit for Social Development*, New Delhi: Discovery Publishing House. ISBN 81-7141-420-6.

Bhaskara Rao, Digumarti, Editor (1999). *International Encyclopaedia of AIDS*, 11 Volumes. New Delhi: Discovery Publishing House. ISBN 81-7141-522-6.

Vol. 1 *Introduction to HIV/AIDS*. ISBN 81-7141-523-7.

Vol. 2 *HIV/AIDS – Issues and Challenges*, 2 Parts. ISBN 81-7141-524-5.

Vol. 3 *HIV/AIDS – Socio Economic Realities*. ISBN 81-7141-524-3.

Vol. 4 *HIV/AIDS – Law Ethics and Human Rights*, 2 Parts. ISBN 81-7141-526-1.

Vol. 5 *AIDS and NGOs*. ISBN 81-7141-527-X.

Vol. 6 *AIDS and Home Care*. ISBN 81-7141-528-8.

Vol. 7 *STD Case Management*. ISBN 81-7141-529-6.

Vol. 8 *HIV/AIDS Prevention and Care – Teaching Modules for Nurses and Midwives*, ISBN 81-7141-530-X.

Vol. 9 *HIV Prevention Education for Educational Institutions*. ISBN 81-7141-531-8.

Vol.10 *Instructional Modules for AIDS Education*. ISBN 81-7141-532-6.

Vol.11 *School Health Education to Prevent AIDS and STD – A Package for Curriculum Planners*. ISBN 81-7141-533-4.

Bhaskara Rao, Digumarti, Editor (2000). *International Encyclopaedia of Human Rights*, 7 Volumes in 13 Parts. New Delhi: Discovery Publishing House. ISBN 81-7141-567-9.

Vol. 1 *International Instruments of Human Rights*, 2 Parts. ISBN 81-7141-569-4.

Vol. 2 *Regional Instruments of Human Rights*. ISBN 81-7141-604-7.

Vol. 3 *Human Rights and the United Nations*, 2 Parts. ISBN 81-7141-605-5.

Vol. 4 *Fact Files of Human Rights*, 3 Parts. ISBN 81-7141-606-3.

Vol. 5 *Study Stories of Human Rights*, 3 Parts. ISBN 81-7141-607-3.

Vol. 6 *International Meetings on Human Rights*, 2 Parts. ISBN 81-714-608-X.

Vol. 7 *Professional Training in Human Rights*. ISBN 81-7141-609-8.

Bhaskara Rao, Digumarti, Editor (2000). *International Encyclopaedia of Science and Technology Education*, 11 Volumes. New Delhi: Discovery Publishing House. ISBN 81-7141-548-2.

Vol. 1 *Science and Technology Education.* ISBN 81-7141-568-7.

Vol. 2 *Science Education in Developing Countries.* ISBN 81-7141-569-9.

Vol. 3 *Organizational Structure of Science.* ISBN 81-7141-570-9.

Vol. 4 *Science Education in Asia and the Pacific.* ISBN 81-7141-571-7.

Vol. 5 *Science and Technology Education For All.* ISBN 81-7141-572-5.

Vol. 6 *Values, Ethics, Talent and Girls in Science and Technology Education.* ISBN 81-7141-573-3.

Vol. 7 *Popularization of Science and Technology Education.* ISBN 81-7141-574-1.

Vol. 8 *Science, Power and Society.* ISBN 81-7141-575-X.

Vol. 9 *Information Technology.* ISBN 81-7141-576-8.

Vol.10 *Teacher Training in Science and Technology Education.* ISBN 81-7142-577-6.

Vol. 11 *Teacher Training in Science and Technology: A Curriculum Framework.* ISBN 81-7141-578-4.

Bhaskara Rao, Digumarti, Editor (2000). *Education For All: Achieving the Goal*, 3 Volumes. New Delhi: APH Publishing Corporation. ISBN 81-7648-152-1.

Vol. I *The Global Consensus.* ISBN 81-7648-155-6.

Vol. II *Mid-Decade Review Reports of Regional Seminars.* ISBN 81-7648-154-8.

Vol. III *Issues and Trends.* ISBN 81-7648-155-6.

Bhaskara Rao, Digumarti, Editor (2001). *Nuclear Materials: Issues and Concerns*, 2 Volumes. New Delhi: Discovery Publishing House. ISBN 81-7141-611-X.

Bhaskara Rao, Digumarti, Editor (2001). *Distance Education in Different Countries.* New Delhi: APH Publishing Corporation. ISBN 81-7648-229-3.

Bhaskara Rao, Digumarti, Editor (2001). *Decentralised Management of Education: Management of Education in Panchayati Raj and Municipal Bodies*. New Delhi: Discovery Publishing House. ISBN 81-7141-617-9.

Bhaskara Rao, Digumarti, Editor (2001). *Electrochemistry for Environmental Protection*. New Delhi: Discovery Publishing House. ISBN 81-7141-619-5.

Bhaskara Rao, Digumarti, Editor (2001). *Global Educational Studies*. New Delhi: Discovery Publishing House. ISBN 81-7141-616-0.

Bhaskara Rao, Digumarti, Editor (2001). *Global Synthesis of Educational Assessment*. New Delhi: Discovery Publishing House. ISBN 81-7141-613-6.

Bhaskara Rao, Digumarti, Editor (2001). *Jomtein Decade of Education*, New Delhi: Discovery Publishing House. ISBN 81-7141-618-7.

Bhaskara Rao, Digumarti, Editor (2001). *World Conference on Education for All*. New Delhi: APH Publishing Corporation. ISBN 81-7141-274-9.

Bhaskara Rao, Digumarti, Editor (2001). *World Conference on Higher Education*. New Delhi: Discovery Publishing House. ISBN 81-7141-610-1.

Bhaskara Rao, Digumarti, Editor (2001). *World Conference on Science*. New Delhi: Discovery Publishing House. ISBN 81-7141-612-8.

Bhaskara Rao, Digumarti, Editor (2003). *Inspiring Experiences in Teacher Education*. New Delhi: Discovery Publishing House. ISBN 81-7141-656-X.

Bhaskara Rao, Digumarti, Editor (2003). *International Studies in Education*, 3 Volumes. New Delhi: Discovery Publishing House. ISBN 81-7141-647-0.

Bhaskara Rao, Digumarti, Editor (2003). *Military Conversion: Impact on Science and Technology*. New Delhi: Discovery Publishing House. ISBN 81-7141-578-4.

Bhaskara Rao, Digumarti, Editor (2003). *United Nations Millennium Summit*. New Delhi: Discovery Publishing House. ISBN 81-7141-632-2.

Bhaskara Rao, Digumarti, Editor (2003). *World Assembly on Aging*. New Delhi: Discovery Publishing House. ISBN 81-7141-637-3.

Bhaskara Rao, Digumarti, Editor (2003). *World Conference on Human Rights*. New Delhi: Discovery Publishing House. ISBN 81-7141-661-6.

Bhaskara Rao, Digumarti, Editor (2003). *World Education Forum*. New Delhi: Discovery Publishing House. ISBN 81-7141-639-X.

Bhaskara Rao, Digumarti, Editor (2003). *Education, Employment and Human Resource Development*. New Delhi: Discovery Publishing House. ISBN 81-7141-681-0.

Bhaskara Rao, Digumarti, Editor (2003). *Successful Schooling*. New Delhi: Discovery Publishing House. ISBN 81-7141-677-2.

Bhaskara Rao, Digumarti, Editor (2003). *European Education and Teachers*. New Delhi: Discovery Publishing House. ISBN 81-7141-702-7.

Bhaskara Rao, Digumarti, Editor (2003). *Teachers in a Changing World*. New Delhi: Discovery Publishing House. ISBN 81-7141-694-2.

Bhaskara Rao, Digumarti, Editor (2004). *International Guidelines on Open and Distance Teacher Education*. New Delhi: Discovery Publishing House. ISBN 81-7141-777-9.

Bhaskara Rao, Digumarti, Editor (2004). *Adult Learning in the 21st Century*. New Delhi: Discovery Publishing House. ISBN 81-7141-797-3.

Bhaskara Rao, Digumarti, Editor (2004). *Educational Practices: Research and Recommendations*. New Delhi: Discovery Publishing House. ISBN 81-7141-835-X.

Bhaskara Rao, Digumarti, Editor (2004). *General Secondary Education In the 21st Century*. New Delhi: Discovery Publishing House.

Bhaskara Rao, Digumarti, Editor (2004). *International Encyclopaedia of Learning to Live Together*, 4 Volumes. New Delhi: Discovery Publishing House. ISBN 81-7141-848-1.

Vol. 1 International Conference on Learning to Live Together.

Vol. 2 Globalization and Living Together.

Vol. 3 Curriculum for Learning to Live Together.

Vol. 4 Science Education for the Contemporary Society.

Bhaskara Rao, Digumarti, Editor (2004). *Reforming Secondary Education*. New Delhi: Discovery Publishing House. ISBN 81-7141-843-0.

Bhaskara Rao, Digumarti, Editor (2004). *Human Rights Education*. New Delhi: Discovery Publishing House. ISBN 81-7141-882-1.

Bhaskara Rao, Digumarti, Editor (2004). *United Nations Decade for Human Rights Education*. New Delhi: Discovery Publishing House. ISBN 81-7141-887-2.

Bhaskara Rao, Digumarti, Editor (2004). *Technical and Vocational Education and Training in the 21st Century*. New Delhi: Discovery Publishing House. ISBN 81-7141-984-4.

Bhaskara Rao, Digumarti, Editor (2005). *Encyclopaedia of Education For All*, 5 Volumes. New Delhi: Discovery Publishing House.

Bhaskara Rao, Digumarti and B.S.V. Dutt, Editors (2003). *Education: Programmes and Policies*. New Delhi: APH Publishing Corporation. ISBN 81-7648-470-9.

Bhaskara Rao, Digumarti, C.A.P. Swamy and B.S.V. Dutt (1997). *Self-Evaluation in Student Teaching*. New Delhi: Discovery Publishing House. ISBN 81-7141-374-9.

Bhaskara Rao, Digumarti and D. Naresh Kumar (2004). *School Teacher Effectiveness*. New Delhi: Discovery Publishing House. ISBN 81-7141-782-5.

Bhaskara Rao, Digumarti and D. Sridhar (2002). *Job Satisfaction of School Teachers*. New Delhi: Discovery Publishing House. ISBN 81-7141-652-7.

Bhaskara Rao, Digumarti, C. Sridevi and K. Vijaya (1995). *Achievement in Social Studies*. New Delhi: Discovery Publishing House. ISBN 81-7141-281-5.

Bhaskara Rao, Digumarti and Digumarti Pushpa Latha (1994). *Achievement in Biology*. New Delhi: Discovery Publishing House. ISBN 81-7141-264-5.

Bhaskara Rao, Digumarti and Digumarti Pushpa Latha (1995). *Achievement in English*. New Delhi: Discovery Publishing House. ISBN 81-7141-283-1.

Bhaskara Rao, Digumarti and Digumarti Pushpa Latha (1994). *Achievement in Science*. New Delhi: Discovery Publishing House. ISBN 81-7141-280-7.

Bhaskara Rao, Digumarti and Digumarti Pushpa Latha (1995). *Achievement in Mathematics*. New Delhi: Discovery Publishing House. ISBN 81-7141-278-5.

Bhaskara Rao, Digumarti and Digumarti Pushpa Latha (2004). *Education for Women*. New Delhi: Discovery Publishing House. ISBN 81-7141-873-2.

Bhaskara Rao, Digumarti, Digumarti Pushpa Latha and Digumarthi Harshitha, Editors (2001). *Biological Warfare*. New Delhi: Discovery Publishing House. ISBN 81-7141-597-0.

Bhaskara Rao, Digumarti, Digumarti Pushpa Latha and Digumarthi Harshitha, Editors (2001). *Women as Educators*. New Delhi: Discovery Publishing House. ISBN 81-7141-602-0.

Bhaskara Rao, Digumarti and Digumarthi Harshitha (2004). *Adjustment of Adolescents*. New Delhi: APH Publishing House. ISBN 81-7648-836-8.

Bhaskara Rao, Digumarti and Digumarthi Harshitha, Editors (2001). *Education in India*. New Delhi: APH Publishing House. ISBN 81-7648-207-2.

Bhaskara Rao, Digumarti and Digumarti Pushpa Latha, Editors (1998). *International Encyclopaedia of Women*, 5 Volumes. New Delhi: Discovery Publishing House. ISBN 81-7141-410-9 (set).

Vol. 1 *Status of World's Women*. ISBN 81-7141-494-X.

Vol. 2 *Women, Education and Empowerment*. ISBN 81-7141-498-1.

Vol. 3 *Women Challenges and Advancement*. ISBN 81-7141-497-4.

Vol. 4 *Women and Family Health*. ISBN 81-7141-497-4.

Vol. 5 *Women and International Action*. ISBN 81-7141-498-2.

Bhaskara Rao, Digumarti, Digumarti Pushpa Latha and Digumarthi Harshitha, Editors (2001). *Assessing Learning Achievement*. New Delhi: Discovery Publishing House. ISBN 81-7141-601-2.

Bhaskara Rao, Digumarti, Digumarti Pushpa Latha and Digumarthi Harshitha, Editors (2001). *Energy Security*. New Delhi: Discovery Publishing House. ISBN 81-7141-598-9.

Bhaskara Rao, Digumarti, Digumarthi Harshitha and K.R.S. Sambasiva Rao, Editors (1999). *Advanced Biotechnology*. New Delhi: Discovery Publishing House. ISBN 81-7141-516-4.

Bhaskara Rao, Digumarti and K.R.S. Sambasiva Rao, Editors (1996). *Current Trends in Indian Education*. New Delhi: Discovery Publishing House. ISBN 81-7141-311-0.

Bhaskara Rao, Digumarti and E. Sreekanth Babu (2004). *Educational Interests of School Students*. New Delhi: Discovery Publishing House. ISBN 81-7141-837-6.

Bhaskara Rao, Digumarti and K. Vijaya (1995). *A Text Book Evaluation*. Ambala Cantt: The Associated Publishers.

Bhaskara Rao, Digumarti and M.A. Fayaz (2004). *Problems of Primary School Drop-outs*. New Delhi: Discovery Publishing House. ISBN 81-7141- 834-1.

Bhaskara Rao, Digumarti and N.V.M. Mohana Rao (2002). *Problems of Mentally Handicapped Children*. New Delhi: Discovery Publishing House. ISBN 81-7141-645-4.

Bhaskara Rao, Digumarti and S. Chandra Mohan (2002). *Sports Management*. New Delhi: APH Publishing House. ISBN 81-7648-467-9.

Bhaskara Rao, Digumarti and S.A. Khader (2004). *Problems of Private School Teachers*. New Delhi: Discovery Publishing Corporation. ISBN 81-7141-838-4.

Bhaskara Rao, Digumarti and S.A. Khader (2004). *School Education in India*. New Delhi: Discovery Publishing Corporation. ISBN 81-7141-849-X.

Bhaskara Rao, Digumarti and Sk. Johni Basha (2004). *Teachers' Population Education Awareness*. New Delhi: Discovery Publishing House. ISBN 81-7141-832-5.

Bhaskara Rao, Digumarti, V.V. Rao, V.V. Lakshmi and V.V. Krishna, Editors (1999). *Status and Advancement of Women*. New Delhi: APH Publishing Corporation. ISBN 81-7648-169-6.

Appala Naidu, P.Ch., Author and Digumarti Bhaskara Rao, Editor (2007). *Feedback Methods and Student Performance*. New Delhi: Discovery Publishing House. ISBN 81-8356-284-1.

Babu, P.C., Author and Digumarti Bhaskara Rao, Editor (2004). *Flowers of Wisdom*. New Delhi: Discovery Publishing House. ISBN 81-7141-695-0.

Bujji Babu, K., Author and Digumarti Bhaskara Rao, Editor (2007). *Teaching Aptitude of Primary School Teachers*. New Delhi: Sonali Publications. ISBN 81-8411-083-9.

Amala, P.A. and Anupama, P., Authors and Digumarti Bhaskara Rao, Editor (2004). *History of Education*. New Delhi: Discovery Publishing House. ISBN 81-7141-860-0.

Bhagya Lakshmi, L., Author and Digumarti Bhaskara Rao, Editor (2000). *Reading and Comprehension*. New Delhi: Discovery Publishing House. ISBN 81-7141-543-1.

Bhasha, S.A., Author and Digumarti Bhaskara Rao, Editor (2004). *Methods of Teaching Geography*. New Delhi: Discovery Publishing House. ISBN 81-7141-807-4.

Bhuvaneswara Lakshmi, Gadde, Author and Digumarti Bhaskara Rao, Editor(2000). Attitude Towards Science. New Delhi: Discovery Publishing House. ISBN 81-7141-541-6.

Bhuvaneswara Lakshmi, G., Author and Digumarti Bhaskara Rao, Editor (2004). *Methods of Teaching Life Science*. New Delhi: Discovery Publishing House. ISBN 81-7141-804-X.

Bhuvaneswara Lakshmi, G. and K. Subba Rao, Authors and Digumarti Bhaskara Rao, Editor (2004). *Methods of Teaching Biology*. New Delhi: Discovery Publishing House. ISBN 81-7141-914-3.

Chary, K.V.N.B., Author and Digumarti Bhaskara Rao, Editor (2006). *Techniques of Teaching Physics*. New Delhi: Sonali Publications. ISBN 81-8411-046-4.

Chowdary, S.B.J.R. and Naga Raju, Authors and Digumarti Bhaskara Rao, Editor (2004). *Mastery of Teaching Skills*. New Delhi: Discovery Publishing House. ISBN 81-7141-861-9.

Dayakara Reddy, V. and Digumarti Bhaskara Rao, Editors (2006). *Value-Oriented Education*. New Delhi: Discovery Publishing House. ISBN 81-8356-051-2.

Devraj, T.A.S., Author and Digumarti Bhaskara Rao, Editor (1997). *Trace Analysis of Uranium and Thorium*. New Delhi: Discovery Publishing House. ISBN 81-7141-375-7.

Durga Rani, K., Author and Digumarti Bhaskara Rao, Editor (2000). *Educational Aspirations and Scientific Attitudes*. New Delhi: Discovery Publishing House. ISBN 81-7141-555-5.

Dutt, B.S.V. and Digumarti Bhaskara Rao (2001). *Empowering Primary Teachers*. New Delhi: Discovery Publishing House. ISBN 81-7141-615-2.

Dutt, B.S.V., Author and Digumarti Bhaskara Rao, Editor (2004). *Comparative Education*. New Delhi: Discovery Publishing House. ISBN 81-7141-912-7.

Ediger, Marlow and Digumarti Bhaskara Rao (1996). *Science Curriculum*. New Delhi: Discovery Publishing House. ISBN 81-7141-321-8.

Ediger, Marlow and Digumarti Bhaskara Rao (2000). *Teaching Mathematics Successfully*. New Delhi: Discovery Publishing House. ISBN 81-7141-552-0.

Ediger, Marlow and Digumarti Bhaskara Rao (2001). *Teaching Science Successfully*. New Delhi: Discovery Publishing House. ISBN 81-7141-600-4.

Ediger, Marlow and Digumarti Bhaskara Rao (2001). *Teaching Social Studies Successfully*. New Delhi: Discovery Publishing House. ISBN 81-7141-596-2.

Ediger, Marlow and Digumarti Bhaskara Rao (2002). *Philosophy and Curriculum*. New Delhi: Discovery Publishing House. ISBN 81-7141-631-4.

Ediger, Marlow and Digumarti Bhaskara Rao (2002). *Improving School Administration*. New Delhi: Discovery Publishing House. ISBN 81-7141-633-0.

Ediger, Marlow and Digumarti Bhaskara Rao (2002). *Elementary Curriculum*. New Delhi: Discovery Publishing House. ISBN 81-7141-658-6.

Ediger, Marlow and Digumarti Bhaskara Rao (2003). *Language Arts Curriculum*. New Delhi: Discovery Publishing House. ISBN 81-7141-657-8.

Ediger, Marlow and Digumarti Bhaskara Rao (2003). *Psychology and Curriculum*. New Delhi: Discovery Publishing House. ISBN 81-7141-691-8.

Ediger, Marlow and Digumarti Bhaskara Rao (2003). *Teaching Language Arts Successfully*. New Delhi: Discovery Publishing House. ISBN 81-7141-678-0.

Ediger, Marlow and Digumarti Bhaskara Rao (2003). *School Curriculum and Administration*. New Delhi: Discovery Publishing House. ISBN 81-7141-709-4.

Ediger, Marlow and Digumarti Bhaskara Rao (2003). *Teaching Mathematics in Elementary Schools*. New Delhi: Discovery Publishing House. ISBN 81-7141-687-X.

Ediger, Marlow and Digumarti Bhaskara Rao (2003). *Teaching Science in Elementary Schools*. New Delhi: Discovery Publishing House. ISBN 81-7141-698-5.

Ediger, Marlow and Digumarti Bhaskara Rao (2003). *School Curriculum and Administration*. New Delhi: Discovery Publishing House. ISBN 81-7141-709-4.

Ediger, Marlow and Digumarti Bhaskara Rao (2003). *Elementary Curriculum Improvement*. New Delhi: Discovery Publishing House. ISBN 81-7141-740-X.

Ediger, Marlow and Digumarti Bhaskara Rao (2004). *Modern Elementary School*. New Delhi: Discovery Publishing House.

Ediger, Marlow and Digumarti Bhaskara Rao (2004). *School Organisation*. New Delhi: Discovery Publishing House. ISBN 81-7141-843-0.

Ediger, Marlow and Digumarti Bhaskara Rao (2004). *Relevancy in Elementary Curriculum*. New Delhi: Discovery Publishing House. ISBN 81-7141-845-9.

Ediger, Marlow and Digumarti Bhaskara Rao (2005). *Quality School Education*. New Delhi: Discovery Publishing House. ISBN 81-8356-022-9.

Ediger, Marlow and Digumarti Bhaskara Rao (2006). *Successful School Education*. New Delhi: Discovery Publishing House. ISBN 81-8356-054-7.

Ediger, Marlow and Digumarti Bhaskara Rao (2006). *Successful School Administration*. New Delhi: Discovery Publishing House. ISBN 81-8356-046-6.

Ediger, Marlow and Digumarti Bhaskara Rao (2006). *Issues in School Curriculum*. New Delhi: Discovery Publishing House. ISBN 81-8356-052-0.

Ediger, Marlow and Digumarti Bhaskara Rao (2006). *Community College – Curriculum and Teaching*. New Delhi: Discovery Publishing House. ISBN 81-8356-053-9.

Ediger, Marlow and Digumarti Bhaskara Rao (2007). *Administration of Schools*. New Delhi: Discovery Publishing House. ISBN 81-8356-244-2.

Ediger, Marlow and Digumarti Bhaskara Rao (2006). *Reading Curriculum and Instruction*. New Delhi: Discovery Publishing House. ISBN 81-8356-266-3.

Ediger, Marlow and Digumarti Bhaskara Rao (2007). *Curriculum Organisation*. New Delhi: Discovery Publishing House. ISBN 81-8356-205-1.

Ediger, Marlow and Digumarti Bhaskara Rao (2007). *Curriculum of School Subjects*. New Delhi: Discovery Publishing House. 81-8356-207-8.

Ediger, Marlow, B.S.V. Dutt and Digumarti Bhaskara Rao (2003). *Teaching English Successfully*. New Delhi: Discovery Publishing House. ISBN 81-7141-707-8.

Elizabeth, M.E.S., Author and Digumarti Bhaskara Rao, Editor (2004). *Methods of Teaching English*. New Delhi: Discovery Publishing House. ISBN 81-7141-809-0.

Elizabeth, M.E.S., Author and Digumarti Bhaskara Rao, Editor (2004). *Acquisition of English Vocabulary*. New Delhi: Discovery Publishing House. ISBN 81-8356-075-X.

Fatima, Sk. Author and Digumarti Bhaskara Rao, Editor (2007). *Reasoning Ability of School Students*. New Delhi: Discovery Publishing House. ISBN 81-8356-330-9.

Gopala Krishna, M., Author and Digumarti Bhaskara Rao, Editor (2007). *Techniques of Teaching Physical Education*. New Delhi: Sonali Publications. ISBN 81-8411-044-8.

Gopala Krishna, M., Author and Digumarti Bhaskara Rao, Editor (2007). *Techniques of Teaching Education*. New Delhi: Sonali Publications. ISBN 81-8411-062-6.

Harshitha, Digumarthi, Author and Digumarti Bhaskara Rao, Editor (2004). *Methods of Teaching Information Technology*. New Delhi: Discovery Publishing House. ISBN 81-7141-805-8.

Harshitha, Digumarthi, Author and Digumarti Bhaskara Rao, editor (2007). *Techniques of Teaching Computer Science*. New Delhi: Sonali Publications. ISBN 81-8411-036-7.

Indira Devi, Author and J. Prasanth Kumar and Digumarti Bhaskara Rao, Editors (2004). *Values in Language Text Books*. New Delhi: Discovery Publishing House. ISBN 81-7141-833-3.

Jalaja Kumari, C., Author and Digumarti Bhaskara Rao, Editor (2004). *Methods of Teaching Educational Technology*. New Delhi: Discovery Publishing House. ISBN 81-7141-810-4.

Jalaja Kumari, C., Author and Digumarti Bhaskara Rao Editor (2007). *Job Satisfaction of Teachers*. New Delhi: Discovery Publishing House. ISBN 81-8356-329-5.

Janardhan Reddy, B., Author and Digumarti Bhaskara Rao, Editor (2006). *Techniques of Teaching Sociology*. New Delhi: Sonali Publications. ISBN 81-8411-042-1.

Jayasree, K., Author and Digumarti Bhaskara Rao, Editor (1999). *Correlates of Socialisation*. New Delhi: Discovery Publishing House. ISBN 81-7141-517-2.

Jayasree, K., Author and Digumarti Bhaskara Rao, Editor (2004). *Methods of Teaching Science*. New Delhi: Discovery Publishing House. ISBN 81-7141-801-5.

John Babu, C., Author and T.J.R. Prasad, G.M. Madhukar and Digumarti Bhaskara Rao, Editors (1996). *Problem Solving in Mathematics*. New Delhi: APH Publishing Corporation. ISBN 81-7648-273-0.

Joseph Raju, B and G.A. Anitha, Authors and Digumarti Bhaskara Rao, Editor (2004). *Population Education*. New Delhi: Sonali Publications. ISBN 81-88836-31-3.

Lalitha, T., Author and K.S. Prabhakaram, D.S.N. Sastry and Digumarti Bhaskara Rao, Editors (2004). *Educational Philosophic Beliefs*. New Delhi: Discovery Publishing House. ISBN 81-7141-765-5.

Krishna, G., Author and Digumarti Bhaskara Rao, Editor (2006). *Techniques of Teaching Physical Education*. New Delhi: Discovery Publishing House. ISBN 81-8411-044-8.

Kumar Raja, G., Author and Digumarti Bhaskara Rao, Editor (2007). *Principles of Primary School*. New Delhi: Sonali Publications. ISBN 81-8411-054-5.

Lakshmi Kumari, V., Author and Digumarti Bhaskara Rao, Editor (2006). *Techniques of Teaching Home Science*. New Delhi: Discovery Publishing House. ISBN 81-8411-048-0.

Madhava, K., Author and Digumarti Bhaskara Rao, Editor (2008). *Personality of Adolescent Students*. New Delhi: Sonali Publications.

Madhu Bala, Jampala, Author and Digumarti Bhaskara Rao, Editor (2004). *Methods of Teaching Exceptional Children*, New Delhi: Discovery Publishing House. ISBN 81-7141-802-3.

Madhu Bala, Jampala, Author and Digumarti Bhaskara Rao, Editor (2007). *Adjustment Problems of Hearing Impaired*. New Delhi: Discovery Publishing House. ISBN 81-7141-831-7.

Marja, Talvi and Digumarti Bhaskara Rao, Editors (1996). *Educational Leadership and Social Changes*, New Delhi: Discovery Publishing House. ISBN 81-7141-320-X.

Mohana Sundari, C., Author and B. Prasad Babu and Digumarti Bhaskara Rao, Editors (2008). *Stress Among Pregnant Women* New Delhi: Discovery Publishing House. ISBN 978-81-8356-316-1.

Mouni Suvarna Raju, T. J., Author and M.V.R.Raju, B. Prasad Babu and Digumarti Bhaskara Rao, Editors (2008). *Personality and Adjustment of Hostel Students*. New Delhi: Sonali Publications.

Naga Kumari, U., Author and Digumarti Bhaskara Rao, Editor (2008). *Science Process Skills of School Students*. New Delhi: Discovery Publishing House. ISBN 978-81-8356-263-8.

Nageswara Rao, S. and M. Srihari, Authors and Digumarti Bhaskara Rao, Editor (2004). *Guidance and Counselling*. New Delhi: Discovery Publishing House. ISBN 81-7141-840-6.

Nageswara Rao, S., Author and Digumarti Bhaskara Rao, Editor (2006). *Techniques of Teaching Psychology*. New Delhi. Discovery Publishing House. ISBN 81-8411-040-5.

Nageswara Rao, S. and P. Sridhar, Authors and Digumarti Bhaskara Rao, Editor (2004). *Methods and Techniques of Teaching*. New Delhi: Sonali Publications. ISBN 81-88836-33-8.

Nirmala Jyothi, M., Author and Digumarti Bhaskara Rao, Editor (2003). *Non-detention System in School Education*. New Delhi: Discovery Publishing House. ISBN 81-7141-654-3.

Padma Tulasi, G., Author and Digumarti Bhaskara Rao, Editor (2004). *Methods of Teaching Elementary Science*. New Delhi: Discovery Publishing House. ISBN 81-7141-871-6.

Pala Prasada Rao, V., Author and K. N. Rani and D. Bhaskara Rao, Editors (2004). *India Pakistan: Partition Perspectives in Indo English Novels*. New Delhi: Discovery Publishing House. ISBN 81-7141-871-6.

Pala Prasada Rao, V., Author and D. Bhaskara Rao, Editors (2008). *Functioning of Autonomous Colleges*. New Delhi: Discovery Publishing House. ISBN 978-81-8356-258-4.

Pitchi Reddy, M., Author and Digumarti Bhaskara Rao, Editor (2007). *Techniques of Teaching Social Sciences*. New Delhi: Sonali Publications. ISBN 81-8411-066-X.

Prasad Babu, B., Author and P. Madhu and Digumarti Bhaskara Rao, Editors (2006). *Psychological Adjustment and Well-being*. New Delhi: Discovery Publishing House. ISBN 81-8356-204-3.

Prasad Babu, B., Author and M.V.R. Raju and Digumarti Bhaskara Rao, Editors (2006). *Behavioural Problems of School Children*. New Delhi: Discovery Publishing House. ISBN 81-8356-206-X.

Prabhakaram, K.S., Author and Digumarti Bhaskara Rao, Editors (1998). *Concept Attainment Model in Mathematics Teaching*. New Delhi: Discovery Publishing House. ISBN 81-7141-424-9.

Prasanth Kumar, J., Author and Digumarti Bhaskara Rao, Editor (1998). *Effectiveness of Distance Education System*. New Delhi: Discovery Publishing House. ISBN 81-7141-437-0.

Prasanth Kumar, J., Author and Digumarti Bhaskara Rao, Editor (2004). *Methods of Teaching Civics*. New Delhi: Discovery Publishing House. ISBN 81-7141-806-6.

Prasanth Kumar, J., Author and G. Sundara Rao and Digumarti Bhaskara Rao, Editors (2000). *Open University Student Support Services*. New Delhi: Discovery Publishing House. ISBN 81-7141-550-4.

Raja Kumari, M.A. and D.R.S. Sundari, Authors and Digumarti Bhaskara Rao, Editor (2004). *Special Education*. New Delhi: Discovery Publishing House. ISBN 81-7141-846-5.

Raja Kumari, M.A. and D.R.S. Sundari, Authors and Digumarti Bhaskara Rao, Editor (2004). *Methods of Teaching Educational Psychology*. New Delhi: Discovery Publishing House. ISBN 81-7141-820-1.

Ramatulasamma, K., Author and Digumarti Bhaskara Rao, Editor (2002). *Job Satisfaction of Teacher Educators*. New Delhi: Discovery Publishing House. ISBN 81-7141-655-1.

Rama Krishnaiah, D., Author and Digumarti Bhaskara Rao, Editor (1998). *Job Satisfaction of College Teachers*. New Delhi: Discovery Publishing House. ISBN 81-7141-438-9.

Rama Kumar Ratnam, M.V., Author and Digumarti Bhaskara Rao, Editor (1998). *Dukkha: Suffering in Early Buddhism*. New Delhi: Discovery Publishing House. ISBN 81-7141-653-5.

Rama Krishna Prasad and P. Vide Sagar, Authors and Digumarti Bhaskara Rao, Editor (2004). *Methods of Teaching Physical Education*. New Delhi: Discovery Publishing House. ISBN 81-7141-868-6.

Rama Seshaiah, P. Author and Digumarti Bhaskara Rao, Editor (2004). *Methods of Teaching Home Science*. New Delhi: Discovery Publishing House. ISBN 81-7141-916-X.

Rama Swamy, K., Author and Digumarti Bhaskara Rao, Editor (2007). *Techniques of Teaching Environmental Science*. New Delhi: Sonali Publications. ISBN 81-8411-035-9.

Ramesh, A.R., Author and Digumarti Bhaskara Rao, Editor (2006). *Techniques of Teaching Commerce*. New Delhi: Sonali Publications. ISBN 81-8411-043-X.

Ramesh, Ghanta and Digumarti Bhaskara Rao, Editors (1998). *Environmental Education: Problems and Prospects*. New Delhi: Discovery Publishing House. ISBN 81-7141-423-0.

Ranga Rao, B., Author and Digumarti Bhaskara Rao, Editor (2007). *Techniques of Teaching Economics*. New Delhi: Sonali Publications. ISBN 81-8411-056-1.

Ranga Rao, R., Author and Digumarti Bhaskara Rao, Editor (2004). *Methods of Teacher Teaching*. New Delhi: Discovery Publishing House. ISBN 81-7141-812-0.

Rajeswari, S. M., Author and T. Santhanam, B. Prasad Babu and Digumarti Bhaskara Rao, Editors (2008). *Stress and Attitude of Women Teachers*. New Delhi: Discovery Publishing House. ISBN 978-81-8356-324-6.

Rani, S.S., Author and Digumarti Bhaskara Rao, Editor (2006). *Techniques of Teaching Botany*. New Delhi: Discovery Publishing House. ISBN 81-8411-037-5.

Rathaiah, Lavu and Digumarti Bhaskara Rao, Editors (1996), *International Innovations in Education*. New Delhi: Discovery Publishing House. ISBN 81-7141-359-5.

Rathaiah, Lavu and Digumarti Bhaskara Rao (1997). *Achievement Correlates*. New Delhi: Discovery Publishing House. ISBN 81-7141-385-4.

Ravi Krishna, M., Author and Digumarti Bhaskara Rao, Editor (2004). *Examination System*. New Delhi: Discovery Publishing House. ISBN 81-7141-824-4.

Ravi Kumar, M., Author and Digumarti Bhaskara Rao, Editor (2004). *Methods of Teaching Computer Science*. New Delhi: Discovery Publishing House. ISBN 81-7141-823-6.

Rudramamba, B., Author and Digumarti Bhaskara Rao, Editor (2003). *Problems of Teaching*. New Delhi: APH Publishing Corporation. ISBN 81-7648-462-8.

Rudramamba, B. and V. Lakshmi Kumari, Authors and Digumarti Bhaskara Rao, Editor (2004). *Methods of Teaching Economics*. New Delhi: Discovery Publishing House. ISBN 81-7141-900-3.

Sambasiva Rao, P., Author and Digumarti Bhaskara Rao, Editor (2007). *Techniques of Teaching Psychology*. New Delhi: Sonali Publications. ISBN 81-8411-040-5.

Sanjeeva Rao, P.C., Author and Digumarti Bhaskara Rao, Editor (1996). *A Text Book of Geology*. New Delhi: Discovery Publishing House. ISBN 81-7141-313-7.

Santhanam, T., B. Prasad Babu and S. Sugandhi, Authors and Digumarti Bhaskara Rao, Editor (2007). *Learning Disabilities and Remedial Programmes*. New Delhi: Discovery Publishing House.

Santhanam, T., B. Prasad Babu and S. Sugandhi, Authors and Digumarti Bhaskara Rao, Editor (2007). *Children with Learning Disabilities*. New Delhi: Sonali Publications.

Sarala, M.M.O., Author and Digumarti Bhaskara Rao, Editor (2006). *Techniques of Teaching English*. New Delhi: Sonali Publications. ISBN 81-8411-047-2.

Satya Narayana, G., Author and Digumarti Bhaskara Rao, Editor (2008). *Attitude Towards Social Studies and Achievement in Social Studies*. New Delhi: Discovery Publishing House. ISBN 978-81-8356-261-4.

Satya Narayana, V., Author and Digumarti Bhaskara Rao, Editor (2001). *Physical Education, Social Attitudes and Leadership Qualities*. New Delhi: Discovery Publishing House. ISBN 81-7141-593-8.

Satya Narayana, P.V.V. and G. Krishna, Authors and Digumarti Bhaskara Rao, Editor (2004). *Curriculum Development and Management*. New Delhi: Discovery Publishing House. ISBN 81-7141-813-9.

Shamsuddin, Sk. and V. Dayakara Reddy, Authors and Digumarti Bhaskara Rao, Editor (2007). *Academic Achievement and Values*. New Delhi: Discovery Publishing House.

Singh, Y.C., Author and Digumarti Bhaskara Rao, Editor (2006). *Techniques of Teaching Science*. New Delhi: Sonali Publications. ISBN 81-8411-041-3.

Sirisha Rani, S., Author and Digumarti Bhaskara Rao, Editor (2007). *Techniques of Teaching Botany*. New Delhi: Sonali Publications. ISBN 81-8411-037-5.

Sivaratnam Reddy, M., Author and Digumarti Bhaskara Rao, Editor (2004). *Creativity in College Students*. New Delhi: Discovery Publishing House. ISBN 81-7141-697-7.

Siva Lakshmi, G.V. and G.L. Subbaiah, Authors and Digumarti Bhaskara Rao, Editor (2004). *Methods of Teaching Environmental Science*. New Delhi: Discovery Publishing House. ISBN 81-7141-839-2.

Srinivas, G., Author and Digumarti Bhaskara Rao, Editor (2007). *Anxiety of Prospective Teachers*. New Delhi: Discovery Publishing House. ISBN 81-8411-084-7.

Srinivas, M. and I. Prasada Rao, Authors and Digumarti Bhaskara Rao, Editor (2004). *Methods of Teaching History*. New Delhi: Discovery Publishing House. ISBN 81-7141-803-1.

Srinivas Rao, P., Author and Digumarti Bhaskara Rao, Editor (2007). *Principles of Secondary School*. New Delhi: Sonali Publications. ISBN 81-8411-058-8.

Srinivasulu Reddy, M. and K.R.S. Sambasiva Rao, Authors and Digumarti Bhaskara Rao, Editor (1999). *A Text Book of Aquaculture*. New Delhi: Discovery Publishing House. ISBN 81-7141-482-6.

Srinivasa Rao, Mandalapu, Author and Digumarti Bhaskara Rao, Editor (2003). *Achievement Motivation and Achievement in Mathematics*. New Delhi: Discovery Publishing House. ISBN 81-7141-674-8.

Srihari, M., Author and Digumarti Bhaskara Rao, Editor (2003). *Values of Prospective Teachers*. New Delhi: Discovery Publishing House. ISBN 81-8356-328-7.

Subba Rao, K., Author and Digumarti Bhaskara Rao, Editor (2007). *School Education Policy*. New Delhi: Discovery Publishing House. ISBN 81-8356-285-X.

Subba Rao, K., Author and Digumarti Bhaskara Rao, Editor (2007). *Education Planning*. New Delhi: Sonali Publications. ISBN 81-8411-053-7.

Sudhakar Reddy, Y., Author and Digumarti Bhaskara Rao, Editor (2003). *Creativity in Adolescents*. New Delhi: Discovery Publishing House. ISBN 81-7141-659-4.

Sunil Kumar, K. and K. Rama Krishana, Authors and Digumarti Bhaskara Rao, Editor (2004). *Methods of Teaching Chemistry*. New Delhi: Discovery Publishing House. ISBN 81-7141-913-5.

Suneetha, G., Author and Digumarti Bhaskara Rao, Editor (2004). *Environmental Awareness of School Students*. New Delhi: Discovery Publishing House.

Sunita, E. and R. Sambasiva Rao, Authors and Digumarti Bhaskara Rao, Editor (2004). *Methods of Teaching Mathematics*. New Delhi: Discovery Publishing House. ISBN 81-7141-915-1.

Surya Madhava, I., Author and Digumarti Bhaskara Rao, Editor (2006). *Techniques of Teaching Geography*. New Delhi: Discovery Publishing House. ISBN 81-8411-034-0.

Surya Madhava, I., Author and Digumarti Bhaskara Rao, Editor (2007). *Techniques of Teaching Political Science*. New Delhi: Discovery Publishing House. ISBN 81-8411-061-8.

Swamy, K.R., Author and Digumarti Bhaskara Rao, Editor (2006). *Techniques of Teaching Environmental Science*. New Delhi: Discovery Publishing House. ISBN 81-8411-035-9.

Swarna Jyothi, K., Author and Digumarti Bhaskara Rao, Editor (2007). *Educational Research*. New Delhi: Sonali Publications. ISBN 81-8411-063-4.

Swarna Latha, C.D., and Digumarti Bhaskara Rao, Editors (2006). *Encyclopaedia of Biotechnology*, 5 Volumes. New Delhi: Discovery Publishing House. ISBN 81-8356-168-3. (set).

Swarupa Rani, T. and J.R. Priyadarshini, Authors and Digumarti Bhaskara Rao, Editor (2004). *Educational Measurement and Evaluation*. New Delhi: Discovery Publishing House. ISBN 81-7141-859-7.

Vanaja, M., Author and Digumarti Bhaskara Rao, Editor (1999). *Inquiry Training Model*. New Delhi: Discovery Publishing House. ISBN 81-7141-515-6.

Vanaja, M., Author and Digumarti Bhaskara Rao, Editor (2004). *Methods of Teaching Physics*. New Delhi: Discovery Publishing House. ISBN 81-7141-867-8.

Valeri V. Koustiouk, Author and Digumarti Bhaskara Rao, Editor (2002). *A Text Book of Cryogenics*. New Delhi: Discovery Publishing House. ISBN 81-7141-642-X.

Vamsi Krishna, V., Author and Digumarti Bhaskara Rao, Editor (2004). *School Psychology*. New Delhi: Discovery Publishing House. ISBN 81-7141-880-5.

Veena Kumari, Balusu and Digumarti Bhaskara Rao (1996). *Operation Black Board*. New Delhi: APH Publishing Corporation. ISBN 81-7024-711-X.

Veena Kumari, Balusu, Author and Digumarti Bhaskara Rao, Editor (2004). *Methods of Teaching Social Studies*. New Delhi: Discovery Publishing House. ISBN 81-7141-899-6.

Veena Kumari, Balusu, Author and Digumarti Bhaskara Rao, Editor (2000). *Psycho-Social Correlates of Achievement*. New Delhi: Discovery Publishing House. ISBN 81-7141-547-4.

Venkata Rao, B., Author and Digumarti Bhaskara Rao, Editor (2007). *Techniques of Teaching Chemistry*. New Delhi: Sonali Publications. ISBN 81-8411-057-X.

Venkata Rao, P. and Digumarti Bhaskara Rao (1989). *A Text Book of Zoology – Junior Intermediate*. Guntur: Vignan Publishers.

Venkata Rao, P. and Digumarti Bhaskara Rao (1989). *A Text Book of Zoology – Senior Intermediate*. Guntur: Vignan Publishers.

Venkateswara Rao, V., Author and Digumarti Bhaskara Rao, Editor (2004). *Problems of Education*. New Delhi: Discovery Publishing House. ISBN 81-7141-841-4.

Venkateswara Rao, V., V. Vijaya Lakshmi and V. Vamsi Krishna, Authors and Digumarti Bhaskara Rao, Editor (2004). *Education For All*. New Delhi: Sonali Publications. ISBN 81-88836-30-3.

Venkateswara Rao, V., V. Vijaya Lakshmi and V. Vamsi Krishna, Authors and Digumarti Bhaskara Rao, editor (2004). *Education in India*. New Delhi: Sonali Publications. ISBN 81-88836-858-9.

Venkateswara Reddy, L. and Narayana, M. L., Authors and Digumarti Bhaskara Rao, Editor (2004). *Education for Dalits*. New Delhi: Discovery Publishing House. ISBN 81-7141-872-4.

Venkateswara Reddy, L. and Narayana, M. L, Authors and Digumarti Bhaskara Rao, Editor (2004). *Methods of Teaching Rural Sociology*. New Delhi: Discovery Publishing House. ISBN 81-7141-811-2.

Venkateswarlu, K. and S.J. Basha, Authors and Digumarti Bhaskara Rao, Editor (2004). *Methods of Teaching Commerce*. New Delhi: Discovery Publishing House. ISBN 81-7141-808-2.

Venugopala Rao, K., Author and Digumarti Bhaskara Rao, Editor (2000). *Teacher Morale in Secondary Schools*. New Delhi: Discovery Publishing House. ISBN 81-7141-551-2.

Venugopala Rao, K., Author and Digumarti Bhaskara Rao, Editor (2007). *Techniques of Teaching History*. New Delhi: Sonali Publications. ISBN 81-8411-059-6.

Vidya, C., Author and Digumarti Bhaskara Rao, Editor (1996). *A Text Book of Nutrition*. New Delhi: Discovery Publishing House. ISBN 81-7141-309-9.

Vimala, T.D., B. Prasad Babu and Digumarti Bhaskara Rao, Editors (2007). *Stress, Coping and Management*. New Delhi: Sonali Publications. ISBN 81-8411-086-3.

Vijaya Bharathi, D., Author and Digumarti Bhaskara Rao, Editor (2000). *Educational Philosophies of Swami Vivekananda and John Dewey*. New Delhi: APH Publishing House. ISBN 81-7648-309-9.

Vijaya Bharathi, D., Author and Digumarti Bhaskara Rao, Editor (2005). *Educational Philosophy of John Dewey*. New Delhi: Discovery Publishing House. ISBN 81-8356-024-5.

Vijaya Bharathi, D., Author and Digumarti Bhaskara Rao, Editor (2005). *Educational Philosophy of Swami Vivekananda*. New Delhi: Discovery Publishing House. ISBN 81-8356-023-7.

Vijaya Lakshmi, D., Author and Digumarti Bhaskara Rao, Editor (2004) *Basic Education*. New Delhi: Discovery Publishing House. ISBN 81-7141-881-3.

Vijaya Lakshmi, V., Author and Digumarti Bhaskara Rao, Editor (2006). *Techniques of Teaching Music*. New Delhi: Discovery Publishing House. ISBN 81-8411-038-3.

Vijaya Kumar, S.J., Author and Digumarti Bhaskara Rao, Editor (2006). *Techniques of Teaching Mathematics*. New Delhi: Sonali Publications. ISBN 81-8411-039-1.

Visalakshi, V., Author and Digumarti Bhaskara Rao, Editor (2006). *Techniques of Teaching Biology*. New Delhi: Sonali Publications. ISBN 81-8411-045-6.

Visalakshi, V., Author and Digumarti Bhaskara Rao, Editor (2007). *Techniques of Teaching Zoology*. New Delhi: Sonali Publications. ISBN 81-8411-055-3.

Bhaskara Rao, Digumarti (1986). *Dhrushya Sravana Bodhanapakaranalu* (Audio Visual Teaching Aids). Guntur: Nagarjuna Publishers.

Bhaskara Rao, Digumarti (1993). *Jeevasashtra Bodhana* (Teaching of Biology). Guntur: Nagarjuna Publishers.

Bhaskara Rao, Digumarti (1995). *Vignanasasthra Bodhana* (Teaching of science) Guntur: Nagarjuna Publishers.

Bhaskara Rao, Digumarti (1997). *Vidya Manovignana Sastram* (Educational Psychology). Guntur: Creative Press.

Bhaskara Rao, Digumarti (1998). *DSC Study Material*. Guntur: Nagarjuna Publishers.

Bhaskara Rao, Digumarti (1998). *Upadhyayudu Vidya*. (Teacher and Education) Guntur: Nagarjuna Publishers.

Bhaskara Rao, Digumarti (1998). *Vidya Drukpadalu* (Perspectives of Education). Guntur: Nagarjuna Publishers.

Bhaskara Rao, Digumarti (1999). *EdCET Teaching Aptitude*. Guntur: Nagarjuna Publishers.

Bhaskara Rao, Digumarti (2001). *Bharata Samajamulo Upadyayudu Vidhya* (Teacher and Education in Emerging Indian Society). Guntur: Sri Nagarjuna Publishers.

Bhaskara Rao, Digumarti (2001). *Bhoutika Sastra Bodhana Padhatulu* (Methods of Teaching Physical Science). Guntur: Sri Nagarjuna Publishers.

Bhaskara Rao, Digumarti (2001). *Jeeva Sastra Bodhana Padhatulu* (Methods of Teaching Biology).Guntur: Sri Nagarjuna Publishers.

Bhaskara Rao, Digumarti (2001). *Vidya Manovignana Sastram* (Educational Psychology). Guntur: Sri Nagarjuna Publishers.

Bhaskara Rao, Digumarti (2003). *Patasala Yajamanyam/Paripalana* (School Management and Administration). Guntur: Sri Nagarjuna Publishers.

Gopala Krishna, G., A. Rama Krishna, K. Subba Rao and Bhaskara Rao, Digumarti (2004). *Jeevasashtra Bodhana Padhatulu* (Methods of Teaching of Biological science). Guntur: Sri Nagarjuna Publishers.

Krishna Murthy, V., K.S. Sudheer Reddy and Digumarti Bhaskara Rao (2004). *Vidya Manovignana Sastra Adharalu* (Foundations of Educational Psychology). Guntur: Sri Nagarjuna Publishers.

Lalini, V., V. Dayakara Reddy, M. Srihari and Digumarti Bhaskara Rao (2004). *Vidya Adharalu* (Foundations of Education). Guntur: Sri Nagarjuna Publishers.

Subba Rao, K.P., P. Ayodhya and Digumarti Bhaskara Rao (2004). *Patasala Yajamanyam – Vidhya Vyavasthalu* (School Management and Systems of Education). Guntur: Sri Nagarjuna Publishers.

Sudhakar, V., B. Ravindra Babu, D.S. Kumar and Digumarti Bhaskara Rao (2004). *Vidya Sanketika Sastram - Computer Vidhya* (Educational Technology and Computer Education). Guntur: Sri Nagarjuna Publishers.

Index

A

Acute stressors, 4
Adjective aspects, 22
Age, 60
Amodio, Toni, 42
Application of learning principles, 24
Appreciation of differences, 24
Armes, 43
Attitude, 14-15, 70-71
– and age, 72-73
– – different management schools, 80-84
– – experience, 74-75
– change, 18
– formation, 15-16
  classical conditioning approach, 15
  genetic influence, 16
  observational learning, 16
  operant conditioning approach, 16
– of teaching, 36
– questionnaire, 85
– towards profession scale, 56-57
Ax, M., 47

B

Basis of effective teaching, 21
Billingsley, 46
Boe, E.E., 48
Brownell, 45

C

Causal factors, 30-33
  administration, 31
  monetary aspects and status, 31-32
  personal problems, 33
  poor working conditions, 31
  pupil's behaviour, 32
  work load, 32
Chen, 49
Cobb, 45
Common acute stressors, 4
Cooper, 45
Cultural improvement, 24

D

Determinants of attitude, 17
  cultural determinants, 17
  psychological determinants, 17
Disillusionment phase, 36
Disorders, 12
  allergy, 12
  gums, 12
  hair loss, 12
  skin disorder, 12
  teeth, 12
Durham, J., 41

E

Effects of stress, 34-36
Emotional control, 23
Experience, 60
External stress, 3

F

Factors influencing attitude, 37-39
  administration, 37
  age, 38-39
  educational level, 39
  monetary benefits, 38
  occupational level, 38
  personal adjustment, 39
  race, 39
  sex, 39
  work group, 38
  working conditions, 38
Findings, 86

G

Gonzalez, P., 47
Griffith, J., 48
Grissmer, 44
Gupta, 41, 42

H

Hastings, 49
Health consequences of stress, 7
  bowel diseases, 9
  diabetes, 11
  eating disorder, 10
  – problems, 10
  gastrointestinal problems, 9
  headaches, 11
  heart diseases, 7
  muscular and joint pain, 11
  pain, 110
  peptic ulcers, 9
  physical effects, 7
  sleep disturbances, 11
  stroke, 11
  susceptibility to infections, 8-9
  weight gain, 10
Hearing impaired children, 69
Henke, R.R., 47
Huling, 43

I

Individualized education program, 26
Internal stress, 3-4
Introduction, 1

K

Keith, F.P., 44
Knowledge of learning principles, 24
Kriby, S.N., 45
Kusheaha, 42
Kyriacou, 42

L

Less experienced teachers, 87
Limitations, 88
Loadmen, 43

M

Mastery of subject matter, 23
Mentally challenged children, 69
Michigan Department of Education, 22
Miller, 48
Moul, 44
Munn, 45

N

National survey, 49
Nichols, 49
Normal children, 63, 65-69

O

Objective of teachers, 35
Objectives of study, 51-52

P

Perkins, 44
Personal attributes, 23
Physical and health status, 23
– problems, 35
Pillay, 41
Psychological effects of stress, 12-14
  concentration, 14
  depression, 13
  learning, 14
  memory, 14
Psychological problems, 35

R

Rawat, 42
Reliability, 58
Responsibility of teachers, 2021
Richardson, 45

S

Sample distribution, 59
Selye, Hans, 2
Sensitivity to differences, 24
Singer, 46
Singh, 42
Socio-cultural factors, 85
Special education, 24-25
  collaboration with other professionals, 26
  direct teaching, 25
  paper work, 26
  responsibility, 25
  working conditions, 27
Special educators, 87
Stress, 2-3, 61
– and age, 71-72
– – different management of schools, 75-80
– – experience, 73-74
– in special education, 33-34
– – teaching, 28-29
  interactive model, 30
  medical model, 29
  professional model, 30
– questionnaire, 55-56, 85
Stressor, 3
Suggestions, 88
Symptoms of stress, 4-7
  acute stress, 4-5
  emotional symptoms, 6-7
  episodic acute stress, 5
  physical symptoms, 5-6

T

Teacher's attitude, 40
Teaching as profession, 18-20
  profession, 18-19
  teaching, 19-20
Troman, 48
Tsuei, 47
Type of children handled, 60, 61
– – school management, 60

U

Understanding human nature and development, 23

V

Validity, 58
Visually challenged children, 63, 65-69

❑❑❑